G000123461

WEST SUSSEX INSTITUTE OF
HIGHER EDUC
WS 2067615 8
AUTHOR
WITHDRAWN
No.
301.54

Housebuilding in Britain's Countryside

Housebuilding in Britain's Countryside

Mark Shucksmith

London and New York

W. SUSSEX INSTITUTE
OF
HIGHER EDUCATION
LIBRARY

First published 1990
by Routledge
11 New Fetter Lane, London EC4P 4EE

Simultaneously published in the USA and Canada
by Routledge
a division of Routledge, Chapman and Hall, Inc.
29 West 35th Street, New York, NY 10001

© 1990 Mark Shucksmith

Laserprinted from author's disks by LaserScript Limited, Mitcham, Surrey
Printed and bound in Great Britain by
Billings & Sons Limited, Worcester

All rights reserved. No part of this book may be reprinted or reproduced or utilised in any form or by any electronic, mechanical, or other means, now known or hereafter invented, including photocopying and recording, or in any information storage or retrieval system, without permission in writing from the publishers.

British Library Cataloguing in Publication Data

Shucksmith, Mark
Housebuilding in Britain's Countryside.
1. Great Britain. Rural regions. Housing
I. Title
363.50941.

ISBN 0-415-04470-7

Library of Congress Cataloging in Publication Data
has been applied for

To Janet, Clare and Anna

Contents

List of figures and tables

Figures

Tables

Acknowledgements

The research described in this book owes much to the encouragement, advice and enthusiasm of Martin Whitby. I am also grateful to my colleagues in the Department of Land Economy, University of Aberdeen, who allowed me the time to complete the thesis from which this book derives, without complaints. Maureen Reid, our departmental secretary, has helped especially. The ideas in this book have also benefited from discussion at many seminars and conferences, both in Britain and abroad, and the contribution of the many participants at these is acknowledged.

Chapter 8 is based on the results of research undertaken jointly with Lynn Watkins, and funded by the University of Aberdeen's Research Committee. I am grateful to Lynn for allowing me to make use of this work.

The foundation of agricultural policy is the house that caught the eye of Mary when she was making up her mind to marry William, who drives the tractor that ploughs the field that grows the corn that feeds the hen that lays the egg that the town eats.

(Young 1943, 12)

The frequency with which references to rural housing have already been made in this chapter reflects the extent to which the housing market acts as an intermediary between the economic changes in agriculture, described earlier in this book, and the kind of social life which is now to be found in the English countryside. This, of course, is using the term housing market in the broadest possible sense to mean the allocation and distribution of housing, including those houses which lie outside the market sector itself, such as local authority housing and tied cottages. How this system of allocation has been operated and controlled has long had a significant effect in moulding the social composition of the rural village, for it has largely influenced who lives where. Housing has therefore been used, either consciously or unconsciously, as an agency of social control.

(Newby 1980, 179)

Chapter one

New housing in the countryside: an introduction

Housing in the countryside is a highly emotive issue at the present time, with media attention and ministerial statements highlighting the delicate balance between the need to protect the British countryside from urban encroachment and the desirability of low-cost housing being made available for village people.

Rural housing is also an issue today because of its significance to other aspects of government policy. Structural surpluses of agricultural production have led to a fundamental rethinking of policies relating to agriculture, the rural economy and land use, and development in the countryside, and this has been perceived by some to signal a loosening of planning controls in the countryside. Concurrently, the government has embarked upon a radical reform of housing policy which stresses the role of the private developer rather than municipal provision: this has added force to the calls of private housebuilders for more greenfield sites, especially in the southeast of England and in the central belt of Scotland. Perhaps most crucially, the country's economic restructuring and industrial revival may be constrained or even threatened by growing regional house price differentials and the associated barriers to labour mobility which have arisen, at least in part, because of the protection of the countryside from new housebuilding.

Inextricably bound up with the debate about countryside protection and whether or not land should be released for housebuilding is the question of who may live in the countryside. Increasingly the British countryside is becoming a gentrified, middle-class countryside, as high house prices and a shortage of low-cost rented accommodation exclude low-and-middle income households. Often this process is characterised as a lack of housing for local village people, reflecting an ideology of localism which tends to obscure divisions of class and status. Nevertheless, the new rural residents have generally been able to adjust their expectations of village life to accommodate increasing social exclusivity and rising property values.

This book addresses both these issues in depth. On the one hand it examines the debate about countryside protection and whether or not land should be released for housebuilding. Can the two goals of policy, countryside preservation and rural housing provision, be reconciled? Second, this book considers who is able to live in the countryside today. It examines differential access to rural housing among competing social groups, and seeks to assess the distributional consequences of policy for these groups. Who gains from existing policies, and who loses? How may less wealthy people be enabled to live in the countryside without a developers' free-for-all?

This introductory chapter is intended to set the scene by explaining the policy context to the discussion of housebuilding in the countryside. It outlines the main elements which have given rise to the current debate, including the agricultural crisis of surplus production, calls for rural diversification and land set-aside; major reforms of housing policy, relying on the cooperation of private property developers and private finance to achieve the government's housing objectives; and macro-economic imperatives towards deregulation, privatisation and reduced public expenditure.

Changing agricultural policies

Agricultural policy in Britain derives largely from the deliberations of the wartime Scott Committee, and was framed in the context of immediate post-war food shortages and rationing. During the 1920s and 1930s agriculture in Britain had suffered a severe depression, with low output prices and consequently low farmland values, due to an absence of government support and the availability of cheap food imports. 'Everywhere land reverted to grass; many farms bore an air of neglect and decay; rural poverty was endemic' (Campbell 1985, 108). With the onset of war, agriculture became a high priority and output rose rapidly. The farmers' contribution to the war effort thus provided them with 'a moral account on which they were able to draw heavily when the war had ended' (Bowers and Cheshire 1983, 59), as well as bolstering the strategic argument for farm support. All parties became committed to permanent agricultural support.

The 1947 Agriculture Act therefore contained a commitment to ensure proper remuneration of farmers and farmworkers and an adequate return on invested capital in agriculture. This was achieved initially through fixed prices paid by government to farmers, then from 1953 by a system of deficiency payments to farmers which allowed consumers to buy food at world prices while farmers received an additional subsidy from government to maintain their guaranteed prices. In addition, the 1950s saw a plethora of capital grants schemes introduced to encourage farmers to increase output through capital investment and capital–labour substitution. By 1960 such payments constituted nearly 40 per cent of all public expenditure in agriculture (Campbell 1985), and this rapid mechanisation allowed output to expand sharply despite a reduction in the cultivated area. As world food prices fell in the 1960s, efforts were made to contain the spiralling costs of deficiency payments through the imposition of 'standard quantities', for example. In essence this system remained in operation until accession to the EC.

The objectives of the EC's Common Agricultural Policy (CAP) are very similar to those of the Agriculture Act 1947, and so entry to the EC in 1973 brought no fundamental change in the direction of British agricultural policy. However the system of supporting farm incomes differed. In highly simplified terms, deficiency payments made by government were replaced by higher food prices paid by consumers: prices are maintained at high levels by intervention – buying of any surplus production by the EC and by variable levies on food imports. So long as the EC remained a net importer, such policies could, in principle, be self-financing with the income from variable levies on imports offsetting the expenditure on surplus storage and disposal, and with the major cost of farm support borne by consumers.

Following entry to the EC, guaranteed prices tended to be higher and the incomes of British farmers rose sharply until the end of the 1970s (Bowers and Cheshire 1983). During this period the British government maintained its commitment to the support of farming in two White Papers (*Food From Our Own Resources* 1975; *Farming and the Nation* 1979), stating that 'a continuing expansion of food production in Britain will be in the national interest'. Yet, even in 1979, agricultural economists were pointing to 'urgent and seemingly intractable problems' (Ashton *et al.* 1979, 1) arising from the rapidly rising costs of financing the CAP, as internal prices rose further above world prices and output increasingly outstripped consumption. Since 1973, average grain production had risen by 1.3 per cent and average dairy production by 2.0 per cent each year up to 1984, while consumption had altered little. The EC has become self-sufficient in all temperate commodities except sheep meat and oilseeds. It is often argued that such effects are endemic: continuous technological change shifts the supply curve outwards while the demand for food remains price and income inelastic, leading to downward pressure on output prices, and inexorably raising the costs of artificially supporting farm prices. There is then a structural tendency to financial crisis which suggests that eventually such policies must be modified by budgetary constraints.

Since 1979, the EC members have attempted various strategies to contain farm spending in order to avoid budgetary crisis. At first they were largely ineffective (Tangermann 1984), but in 1988 a package of reforms was agreed which combined a system of price 'stabilisers' with physical controls on some commodities and measures intended to encourage diversification of income sources. As a result, in 1988 for the first time the share of EC expenditure taken by price support has fallen to under 60 per cent, although this element still dominates the EC budget.

As the revenue available to farmers is diminished, either through stabilisers (essentially price cuts and co-responsibility levies) or through quantitative restrictions, it is expected that resources will leave agriculture. By a curious logic, examined by Lowe and Winter (1987), this expectation that resources will leave agriculture has been presented in Britain in terms of land surplus to agriculture, rather than in terms of over-capitalisation. Several studies in recent years have attempted to estimate the quantity of surplus land that exists, based on the amount of land needed to remain just self-sufficient in basic foodstuffs. Estimates of surplus land in Britain by the year 2000 range from 2.6m hectares (Gould 1986) through 4m hectares (Edwards 1986) to 5.5m hectares in 2015 (North 1987), while CAS (1986) estimates that under different policy scenarios of free-trade, quotas or co-responsibility levies the amount of surplus land would be 22 per cent, 18 per cent or 12 per cent over that needed for self-sufficiency.

In Lowe's view 'talk of a land surplus is a contrivance'. If all support through the CAP were withdrawn, he argues, it is unlikely that there would be any great surplus of land, since a reduction in capital investment would also occur. However, 'such a radical solution to food surpluses finds little favour in agricultural circles' (Lowe 1988, 37) because of the threat it poses not only to farmers' incomes but also to agrochemical input suppliers and to land values.

Nevertheless, government policy rests on an acceptance of the notion of surplus agricultural land. In February 1987, the Agriculture Minister announced that

> with the EC now producing surpluses in many of the main agricultural commodities, a new balance of policies has to be struck, with less support for expanding production; more attention to the demands of the market; more encouragement for alternative uses of the land; more response to the claims of the environment; and more diversity on farms and in the rural economy.

To these ends he announced a £25m package involving annual payments to farmers to plant farm woodlands, an expansion of afforestation, greater support for farmers pursuing environmental objectives in Environmentally Sensitive Areas (ESAs), support for farm enterprise diversification, and research on new crops and livestock. The intention, he explained, was to depress agricultural output without depressing the agricultural economy.

A major element of the strategy was the promotion of the orderly withdrawal of land from production, for example, through the farm woodland scheme, a set-aside scheme and the encouragement of alternative land uses, such as recreational uses. Simultaneously with the Minister of Agriculture's announcement, the Secretary of State for the Environment announced that, because of 'the need to foster the diversification of the rural economy to open up wider and more varied employment opportunities', the longstanding presumption against the development of agricultural land, for all but the best quality land, was to be relaxed.

The set of publications released in 1987 under the general heading 'Farming and Rural Enterprise' emphasised the need to foster the diversification of the rural economy, and noted 'the enormous gains in the efficiency and productivity of our agricultural industry, which means that less land is needed to meet our requirements' (DoE 1988b, 5). Predictably, the government statement that less farmland will be needed in the future excited the interest of several groups, each of which has its own ideas of what should be done with the surplus land, whether it be a 'developers' free-for-all', large-scale afforestation, or farmers turned conservationists. In particular, the volume housebuilders have been able to lodge planning applications with their hopes raised now that land no longer has to be reserved for agriculture: such proposals may now be treated on their own merits in relation not only to agriculture but 'to the need to promote economic activity that provides jobs... and to the need to protect green belts, national parks, AONBs and other areas of good countryside that should be conserved and protected from discordant development' (DoE 1987). While ministers played down the signifi- cance of these changes, the Council for the Protection of Rural England rejected this interpretation, suggesting that local authorities trying to protect the countryside would be placed at the mercy of predatory developers.

Reforms in housing policy

While 'housing in rural areas has to be seen in relation to the changes that are taking place in agriculture and the rural economy' (DoE 1988b, 1), it must also be seen in the context of the equally radical reforms which have been implemented in housing policy. For housing provision has altered considerably in its nature, type and location since the war, and during the 1980s housing policy has been subject to particularly fundamental reform.

Until the 1919 'Addison Act' housing provision was almost entirely the responsibility of private developers and private landlords, who failed to provide decent low-cost housing (Kemp 1989). However, after that date successive governments initiated a series of measures authorising and financing municipal intervention, both in providing houses and in relation to slum clearance and redevelopment. During the 1920s, local authorities concentrated on addressing the housing shortage through the construction of general needs council housing, and then after 1933 their role shifted towards slum clearance and redevelopment, leaving general needs provision to the private sector. Still, 'local authorities played a key role in the implementation of policy throughout the period, within the framework of objectives established by central government' (Malpass and Murie 1987, 66) and there was no direct promotion of home ownership.

At the end of World War II, there was again a serious housing shortage coupled with a daunting legacy of slum conditions. For most of the next thirty years there was general agreement between the political parties that there was a need for high levels of housing construction, at least until 1968, and it was also accepted that there was a significant role for public housing in meeting these needs. Proportionately, the local authorities built at a much higher level than prior to World War II, building over 2.9 million dwellings in the twenty years after 1945. However, there were different phases within this period: between 1945–51 municipalities built over 80 per cent of all new dwellings, while the proportion fell below 50 per cent between 1954–61 as 'the Conservatives inverted the contributions made by the public and private sectors', before the 1964 Labour government in turn re-emphasised the public sector's role (Malpass and Murie 1987). From 1969, the emphasis shifted from new building to rehabilitation and improvement for a number of reasons. The local authority still had a strong interventionist role which was central to the implementation of housing policy. By 1979 the public rented sector had grown to encompass 32 per cent of all households in Britain, and 56 per cent in Scotland.

Reservations began to emerge about the municipal role during the 1970s, however:

> The euphoria of the new development phase evaporated sharply, and growing doubts about the impact of the programme began to be widely expressed. The pressures for a more considered approach to housing provision arose from a number of factors. Concern came to be expressed about the nature of the dwelling structures and neighbourhoods created and their evolution over time, and by the 1970s dwellings less than two decades old were already unlettable. Technical problems were, in many areas,

> reinforced by the style of management and the inadequacies of management systems. The external policy environment was also changing, and the case for rehabilitation and inner city investment gained credibility and political support. Such new ideas for housing provision and the increasingly ailing public sector have had to compete for shrinking public investment resources as housing programmes after 1975 became the leading sector in the attempt to reduce public spending.
>
> (Gibb and MacLennan 1986, 281)

This disenchantment and the changing emphasis of policy led to the imposition of controls over local spending. Until the mid-1970s, councils had considerable freedom of action with central government exercising only an indirect influence through subsidy arrangements and global targets for new building. According to Malpass and Murie (1987) 'it was not until the public expenditure crisis of the mid-1970s that local authorities began to be affected individually and directly by a tightening of central government control over capital spending'. Under the system of Housing Investment Programmes (HIPs) introduced in 1977, central government set a maximum amount to be spent by each local authority on capital works, so that the decision about the appropriate level of local government investment is now taken centrally. Since these controls were introduced, councils have been forced to reduce their housing investment drastically and so to curtail their housebuilding programmes.

Under the Labour government of the mid-1970s, cuts in housing investment and controls on council spending decisions were always presented as a temporary but necessary evil. Since 1979, however, the Conservative government has pursued both cuts in expenditure on public sector housing and controls over municipal freedom for their own sake. A reduction in public expenditure was required in accordance with the government's macroeconomic imperatives of approaching a balanced budget and constraining monetary growth; municipal intervention was to be discouraged in pursuit of deregulation, privatisation and competition. On both these grounds, the building and even the ownership of housing by councils was an impediment.

The local authority's role as housing provider and landlord was assaulted in various ways during the 1980s. In the early 1980s, HIP allocations were sharply reduced as housing investment bore the brunt of the government's public expenditure cutbacks. While this was partially offset in rural districts by capital receipts from council house sales, local authorities in England were prevented from reinvesting more than a small proportion of these receipts (50 per cent prior to 1984; 20 per cent since 1985). The effect of these constraints is evident in the decline in the number of new council houses built in England and Wales from 105,000 in 1976 to 67,000 in 1979, and to only 22,000 in 1985. In Scotland there was a similar decline.

At the same time as they suffered this reduction in their ability to build houses, local authorities also lost stock through the mandatory 'right to buy' at a discount given to tenants in the Housing Act 1980. These council house sales were particularly prevalent in rural districts, so that the landlord role of rural municipalities – already small for historical reasons – was further diminished. The possibilities for the privatisation of councils' stock were broadened again by the provisions of the Housing Act 1988 relating to tenants' transfer: council tenants were given the right to opt for a new landlord in the private sector.

The White Paper announcing this legislation made quite explicit the antipathy of the government towards local housing authorities. Councils were to shed their role as providers of housing, and become instead enablers of private sector provision. While they would remain substantial landlords for some time, this role was also to be diminished in the interests of dismantling municipal monopolies. This represents a considerable departure from the position for most of the last sixty years, during which local authorities have had a central role in the implementation of housing policies, usually as a major provider of rented housing and as landlord. Now 'a steady growth of municipal ownership has given way to demunicipalisation' (Malpass and Murie 1987, 101), and the onus is on the private sector.

Alongside these changes in municipal housing it is also pertinent to refer to the growing political support for the extension of home ownership, which has been regarded since the 1950s as the 'natural' tenure. As Malpass and Murie point out,

> as governments in the 1960s and 1970s relied increasingly on a policy of expanding home ownership, the behaviour of the institutions managing that sector moved into sharper focus. The emergent issue here was the extent to which private sector institutions, such as building societies, were prepared to be incorporated into housing policy.
>
> (Malpass and Murie 1987, 94)

This issue is central to the current debate about land release for housing. For government is now reliant upon the cooperation of the private sector not only for achieving an expansion of home ownership but for the success of its entire housing policy. By downgrading the role of local housing authorities, the Conservative government has placed almost all responsibility for the provision of housing, to rent as well as to buy, on private developers and the related private financial institutions. Housing policy thus relies centrally on the cooperation of private housebuilders, and on their willingness and ability to provide appropriate housing: clearly this places them in a very powerful position, verging on a corporatist relationship with central government. From a corporatist perspective, therefore, 'the planning reforms are a necessary price to pay for public policy to be effective' (Rydin 1986, 9), although there may be a counter effect of other incorporated interests such as rural landowners and farmers. Notwithstanding this, Rydin argues that the latter interests have now been supplanted by housebuilders as the dominant incorporated interest.

The demotion of planning, and indeed the very emergence of the land release issue, may therefore be viewed as a necessary response by government to the increased reliance placed upon private builders in their corporate role.

> With the shift from public to private home ownership, different forms of land release are required. Development land now appears from private landbanks and planning permission is acquired by private developers without the involvement of the local authority either as landowner or as future owner of the housing development. To match these changes, the planning system needs to change so that planning permission is readily forthcoming where the developer seeks it.
>
> (Rydin 1986,7)

Equally, of course, the relaxation of planning controls must be related to other aspects of state deregulation and attempts to remove obstacles to business enterprise in accordance with the supply-side emphasis of the government's economic policy. As Bell and Cloke (1989,13) have observed,

> There is certainly evidence in Britain of a gradual deregulation of the post-war planning system. This has included: a reduced role for county-level strategic planning – with further reductions proposed in the current discussions about the future of development plans; reduced planning controls over the development of agricultural land; reduced planning in special areas (as yet unspecified, but seemingly a rural equivalent of urban enterprise zones); and a reduced role for major planning inquiries... Rural researchers will need to take careful note of the interactivities between the deregulation of planning and other strands of privatisation.

Indeed, the 1985 White Paper, *Lifting The Burden*, may be added to this list, proposing as it did a general relaxation of the planning system with a presumption in favour of allowing development unless harm would be caused 'to interests of acknowledged importance'.

Rural housing in context

Ministers have also found other reasons for concern about housing in rural Britain. The Department of Trade and Industry (DTI), for example, has submitted evidence in favour of the building of new towns and villages in the southeast of England, at Foxley Wood in northeast Hampshire and at Stone Bassett in Oxfordshire. Its argument is that the health of the regional and national economy is being prejudiced by the shortage of housing to rent or to buy inexpensively in the southeast, and that this is a consequence of a scarcity of housing land. The House Builders' Federation has claimed that already this shortage, and its inflationary effects, have affected the entire economy through the house price boom, necessitating substantial rises in interest rates in 1989. The former Environment Minister, Michael Heseltine, has placed this issue at the centre of a personal campaign for the succession to the leadership of the Conservative Party, arguing that planning controls should not be relaxed but that development pressures should be diverted away from the southeast through vigorous regional policy measures. The interrelationship between the housing market and the economy, and its importance, is thus widely acknowledged. According to Lock,

> Land for housing is so expensive, and competition for new homes so great, that would-be first-time buyers are finding nothing they can afford. The price of land has in turn led housebuilders to give up catering for first-timers, and they now build instead for higher income groups.... An economic consequence of this phenomenon is that staff are either having to leave their jobs in the south-east and join the overcrowded labour market in the north, or they are having to live further away from their work and spend a great deal of their income (and large chunks of their lives) on trains and coaches. Companies trying to recruit younger people or fill lower paid posts find they cannot.
>
> (Lock 1989)

Lock, like the DTI, is convinced that 'we need land for housing outside the built-up area of Greater London, no matter how much the locals protest'.

Interestingly enough there are also demands from 'locals' for more housing to be built in rural areas, although these are not the same 'locals' to whom Lock refers. As a recent publication from the House Builders' Federation (1988) notes,

> A key problem in today's countryside is the lack of affordable housing. Rising demands for housing from outside the countryside and a lack of firm action to meet housing needs have contributed

> to a shortage of housing opportunities for countryside people. The young, the elderly, the less well-off and those looking for jobs are just some of the groups finding it increasingly difficult to live and work there.

In similar vein, the National Agricultural Centre Rural Trust, backed by the Rural Development Commission and the Housing Corporation, has called for 'village homes for village people' and pointed to 'the incompatability between house prices and local average incomes coupled with the lack of rented housing' (NACRT 1987). The concern is for 'countryside people', or 'village people' or for children of families who have spent their lives in the villages. Moreover, their claims have been endorsed by the Secretary of State for the Environment, who has accepted 'that there are real unmet needs for low cost housing in rural areas' for local people (DoE 1988a). He finds himself in a dilemma, however, confiding to *The Guardian* (17 June 1989) that 'everyone is concerned about the need for more rural housing but everyone is equally concerned to stop further development outside existing urban areas'. No doubt each of these universes is also comprised of locals !

In the rhetoric of politics then, there are two kinds of 'locals'. On the one hand, locals are middle-class owner-occupiers who object to builders providing new houses to meet regional needs; on the other, locals are young or elderly low-income 'country people' who cannot find affordable housing in the area of their birth. Each of these stereotypes has been presented as justification for reducing or increasing housebuilding in the countryside. But who is rural housing really for? Whose demands or needs are to be met by new housebuilding, and whose are denied by the lack of it? It is particularly helpful to distinguish between the regional demand for more houses which results from smaller households and greater prosperity in the south – the preferences of an increasing number of people to live in the countryside while commuting to work or in retirement – and the necessity of low- income people with jobs in rural areas for low-cost housing in reasonable proximity to their work. Each of these claims on rural housing and housing land may be met in quite different ways and in quite different locations, and this is discussed further in Chapter 8.

Perhaps at this juncture, it is sufficient to recognise that these demands, as expressed in crude political sloganising, have different scales attached to them. The champions of village homes for village people want a few low-cost houses built in or adjacent to each village, amounting perhaps to only a few hundred houses in a whole county. In Gloucestershire, for example, the Country Land-owners Association believes that 2,000 houses over ten years, on 400 half-acre plots would be sufficient (Dunipace 1988). Meeting these needs requires only a few fields here and there. Even at national level, a recent report to the Rural Development Commission has (rather generously) estimated that some 370,000 'affordable village houses' are needed in rural England now, on the basis of parish council estimates, although the NACRT estimates a rather lower current need of 60–100,000. Although these calculations are highly imprecise, and vast in relation to current building programmes of social housing, the scale of land required is small. The NACRT figure implies a requirement throughout England of perhaps 5,000 hectares, and even the high estimate submitted to the Rural Development Commission would necessitate only around 18,000 hectares to meet 'village needs'.

Meeting regional needs or national needs is thought to require far more: the House Builders' Federation quotes a government estimate of 2 million more homes needed in Britain by the end of the century, and this is thought to require land release of 75,000 hectares, equivalent to half the area of South Yorkshire (Lean 1989). This is still very small in relation to the estimates of surplus farmland cited earlier, which vary between around two and five million hectares. Under a scenario of co-responsibility levies, which are broadly similar to the stabilisers now in force in Europe, the CAS (1986) has concluded that the areas with most surplus agricultural land will be East Anglia, the southeast and the southwest. The implication is that more land for housing may be available, subject to planning decisions, outside green belts and other specially protected areas in the very countryside where the greatest demand for housing occurs. Whether and how this surplus land might be released for housing is discussed in some detail in Chapters 7 and 8.

At present, the Government is attempting to walk a tightrope between the demands of house builders for more land and the demands of Conservative voters in the shire counties for the protection of their environment, while also mindful of its green image and the 'village homes' lobby. The balancing act involves headline-making ministerial decisions against high-profile new settlements, while county and local plans are less noticeably amended to ensure more land is zoned for housebuilding. The current slump induced by high interest rates assists in that it has temporarily alleviated pressure for development, especially in the southeast. Village needs are acknowledged in a well-publicised programme of building by village housing associations, which critics suggest is so inadequate in scale that it must be regarded as essentially cosmetic. Again, these policies are reviewed in Chapters 7 and 8.

The structure of this book

The next four chapters probe more deeply into the processes at work in the countryside, and suggest ways of approaching and identifying these forces. Chapter 2 is the most abstract, discussing the theoretical issues raised in this book, and drawing on various methods applied by researchers into urban housing as well as the methods used in earlier rural housing studies. Readers who are less interested in theoretical issues may wish to skip pages 17–28 and read only the final section.

Chapter 3 addresses the first of the two principal themes of the book, that of the conflict of objectives underlying intervention in rural housing markets; and to that end it summarises post-war rural planning policy in relation to housing in the countryside. Chapter 4 turns to the second theme, relating to who can afford to live in the countryside, beginning with a discussion of inequity and need, and then proposing a Weberian typology of rural housing classes. In Chapter 5 this typology forms the framework for a consideration of how access to rural housing resources differs between social groups and for the identification of the most disadvantaged groups.

Chapter 6 is a case study of a controversial attempt by a planning authority to intervene in a rural housing market in order to resolve the efficiency conflict between landscape preservation and housing provision objectives and to influence distributional outcomes in favour of local people. This case study allows the two themes of this book to be elaborated in more detail and in relation to one another. It analyses the policies of the planning authority and their distributional consequences at length, and assesses the motives behind the introduction of its 'locals only' policy from a perspective which seeks to identify the wider constraints under which local 'managers' attempt to mediate between central government and the local population, and between the private sector and social needs.

The next two chapters pursue the two themes of the book in the light of the issues raised in this case study, focusing particularly on the constraints which limit the freedom of action of local housing managers. Chapter 7 considers what alternative policies might be adopted by local authorities, with or without the support of central government, to adjust the balance between competing claims on rural land, while also seeking to improve housing opportunities for presently disadvantaged groups. In particular, how might it encourage the provision of more low-cost housing to rent without damaging the landscape? Such a question reveals the impotence of local planning authorities, and indeed strikes at the fundamental assumptions of the British planning system. The chapter also assesses the options open to local housing authorities, and analyses the financial and administrative constraints which limit their contribution.

Chapter 8 (written with Lynn Watkins) reviews the political dilemma facing central government as it seeks to devise a planning policy for land release and to relate this to its reform of housing policy. Three alternative future scenarios are assessed in terms of their capacity for reconciling political contradictions and in terms of their distributional effects.

Finally, Chapter 9 summarises the main issues and the principal findings of the book. It draws attention to the shortcomings and contradictions of existing policies, and suggests a more effective strategy for central and local government. In addition, this chapter highlights areas in which information on the operation of rural housing markets is deficient, and, therefore, in which further research is desirable.

Chapter two

Perspectives on rural housing research

Previous rural housing research

Until recently, there had been little research into rural housing in Britain compared to the much greater attention given to urban housing issues. In 1982 Phillips and Williams were still able to lament the 'general neglect of rural housing issues', noting that 'only belatedly has there been an awareness that the real problems may be those related to social and economic access to housing' (Phillips and Williams 1982, 3).

Rogers' (1976) review of rural housing was a turning point in this respect. Rogers (1985b, 87–8) himself looks back on this period

> Prior to the mid-1970s rural housing was only studied when it appeared as a clearly-defined topic. Thus, second homes flowered briefly in the late 1960s–early 1970s (e.g. Bielckus, Rogers and Wibberley 1972) and tied housing rose to prominence in mid-decade (e.g. Gasson 1975) particularly when there was the prompting of legislation.... This late and partial concern amongst researchers has inevitably given a varied, though poorly-related, pattern of research.

He identifies five broad approaches in recent British research:

1. The spatial analysis of rural housing (e.g. G. Clark 1982; Phillips and Williams 1983), tending to be 'more descriptive than analytic'.
2. The social welfare approach, classifying rural dwellers in relation to their access to housing, according to their income, social class, etc. (e.g. Pahl 1966; Dunn, Rawson and Rogers 1981).
3. The political science of rural housing, focusing on the operation of the public housing system (e.g. Newby, Bell, Rose and Saunders 1978) and descriptive geography of decision-making (Phillips and Williams 1982).
4. The economic analysis of rural housing (HIDB 1974a; Shucksmith 1981).

5. A series of local case studies, 'polemical rather than academic in tone', and relying upon 'an eclectic, if somewhat superficial, blend of spatial and statistical analysis, social concern and political comment' (e.g. Winter 1980).

Rogers (1987, 147) takes the view that, following his 1976 article, studies of rural housing

> followed in the pattern of many other aspects of the rural economy in showing a lag of about ten years or so behind the study of the same sector in the urban environment. By the mid 1980s there has been the same accelerated romp through facets of study which were pioneered in an urban framework. Access, social justice, managerialism, political economy approaches – all have by now been summarily touched on and the relevant flavour imparted.

Because work has tended to proceed piecemeal, Rogers (1985b, 87) has characterised rural housing research as 'an issue in search of a focus', but despite this fragmentation he recognises (1987, 150) that the major theme developed during the last ten years has been that of differential access to rural housing. Nevertheless, this theme has not been integrated into any overall structure which relates it adequately to rural economy and society.

Few researchers into rural housing in Britain have discussed their methodological foundations explicitly. In a recent textbook, Phillips and Williams (1984, 15) attempt to draw parallels with geographers' studies of urban housing. They adopt Moseley's (1980) simplification of Robson (1979) to argue that 'in the 1970s studies of urban geography passed through four overlapping phases: quantitative model-building, behaviouralism, managerialism, and political economy'. Moseley argued that rural geography was bogged down in the first two phases, which focused on the sufferers of deprivation, rather than on its producers (i.e. managers and power elites). Phillips and Williams illustrate these different approaches in the context of rural housing:

> A quantitative model-building approach would aim to provide a sophisticated description of the distribution of needs and resources in housing and, through statistical analysis of aggregate data, would suggest some of the variables, such as income levels or demand, associated with this distribution. Behaviouralism, on the other hand, would direct attention away from the aggregate level to the individual householder or aspirant, and would examine such features as housing search processes, residential mobility and changing needs. It would reveal how individuals

> react to constraints in the housing market. By contrast, managerialist and political economy perspectives highlight, respectively, the operation and production of such constraints.
>
> (Phillips and Williams 1983,15)

Of the principal contributions to rural housing research, it will be seen that Dunn, Rawson and Rogers' (1981) work and the SDD (1979) study follow an approach broadly similar to that of managerialism, whereas Newby *et al.* (1978) tend more towards a political economy approach (albeit a Weberian one) in many respects, by considering not only the outcomes of rural housing allocation processes but also the political forces which underlie them. Both Moseley, and Phillips and Williams advocate that rural researchers should increasingly adopt managerialist or political economy approaches, and for this reason, and because of the parallel drawn by Rogers between the approaches of rural housing researchers and of those analysing urban housing ten years earlier, it will be instructive to look more closely at the methodological foundations of urban housing studies.

Readers who are less interested in theoretical issues and whose main concern is with rural housing policies may wish to skip the discussion of urban housing research methods. They may find it useful nevertheless to pick up the discussion again on page 28, when it returns to rural housing research methods and the main concerns of this book.

Research methods in urban housing research

Approaches to urban housing research may be characterised as market oriented, managerialist and political economy approaches (McDowell 1982). The market oriented approaches of social ecology and of urban microeconomics have now fallen out of favour for the most part, and instead, 'recent research has stressed the essentially political nature of housing and its provision' (Robson 1979, 71) emphasising the importance of constraints rather than choice in access to housing, and of class-based conflict rather than consensus as the basis of the system by which housing is produced, allocated and consumed.

Bassett and Short (1980) draw a distinction between the ecological and neo-classical approaches, which (they argue) focus on equilibrium conditions, housing choice and social harmony, and the more recent managerialist and political economy approaches, which focus on disequilibrium conditions, housing constraints and social conflict. They emphasise the relationship of each of these approaches to a wider social theory, as suggested in Table 2.1.

Table 2.1 Four approaches to housing research

Approach	*Social theory*	*Areas of enquiry*
1. Ecological	Human ecology	Spatial patterns of residential form
2. Neo-classical	Neo-classical economies	Utility maximisation, consumer choice
3. Managerialism	Weberian	Gatekeepers, constraints
4. Political economy	Marxist	Housing allocation, reproduction of labour force

Source: adapted from Bassett and Short (1980, 2)

Market oriented approaches

Many writers have rejected the social ecology and urban micro-economic approaches as offering inadequate explanations of housing markets. Both approaches attempt to describe who lives where in cities, and in their simplest versions outline a similar pattern of concentric rings with the rich living in the suburbs and the poor in the inner city. MacDowell (1982) points out that both assume unregulated competition for land use, and argues that each is merely a description of processes unique to urban America at the beginning of this century. In addition, the models of the urban economists (Alonso 1960; Wingo 1961) are static equilibrium models which ignore the durability of houses and the dynamic nature of housing markets (MacDowell 1982). These are also rejected by Bassett and Short (1980) on the grounds that utility-maximisation and the access–space trade-off are unrealistic behavioural assumptions and that these models ignore private property institutions and power structures, failing to model the effects of agents other than individuals, and emphasising individual choice rather than the constraints facing individuals. They also ignore tenure. Notably, the housing economist, MacLennan (1982), also rejects such models, concluding that they have distracted attention from important economic aspects of housing markets, and he advocates instead an applied political economy approach. It is clear, however, that his use of the label 'political economy' is intended to convey something far removed from Bassett and Short's use of the term to denote Marxist analysis. As Newby (1982, 135) has argued, the use of the term 'political economy' as a euphemism for Marxist theory is unfortunate, as well as confusing, since it pre-empts the possibility of constructive debate between Marxist and non-Marxist approaches to political economy.

MacLennan (1982a, 59) criticises the access–space model for its specific and empirically unjustifiable assumptions regarding the objects and processes of housing choice. More fundamentally,

> a continuing commitment to the assumptions of optimising behaviour in simple competitive markets which are frictionless and in equilibrium, results in housing demand theory remaining an unconvincing framework for the theoretical and policy analysis of particular housing market phenomena.

MacLennan's main concern is to refine the behavioural premises of economic modelling through a more sophisticated conception of housing choice, which minimises the significance of prior deductive assumptions. Yet this is not to say that constraints are neglected, as alleged by both Robson (1979) and Bassett and Short (1980). MacLennan distinguishes between preferences (the underlying tastes which exist independently of constraints and which are formalised in the utility function) and choice (the outcome of the interaction of preferences and constraints): a model of housing choice thus necessarily incorporates an analysis of constraints. MacLennan also seeks to broaden the concept of equilibrium (Hahn 1973), and to embrace the possibility of disequilibrium.

Nevertheless, this still leaves outstanding a number of criticisms of economic models: that they assume social harmony rather than conflict, that constraints are often neglected, and that too much emphasis is placed on individuals at the expense of institutions, power structures and class interests. MacLennan accepts that neo-classical economic models may not be appropriate for the empirical analysis of housing markets, and advocates instead 'the applied economic analysis of policy (or political economy)'. This, he argues, implies more than mere description of market behaviour:

> For instance, the analyst must enquire as to what are the objectives of government, what role housing plays in attaining such objectives, who actually makes housing policy, what economic and political constraints restrict action, what advice and information exists and what model of the housing system is used by policy makers?
>
> (MacLennan 1982a, 142)

The applied economic analysis of policy, as defined by MacLennan, thus appears to have moved well away from the neo-classical tradition of Alonso and Wingo to embrace concerns more usually associated with managerialist and political economy approaches. MacLennan seeks both to refine the behavioural premises of urban economic models and also to incorporate wider political constraints and the possibility of conflict. Such an approach, if it can be developed, may contribute to the formulation of a non-Marxist political economy. However, it is not at all clear what theory will inform this proposed broadening of economists' concerns, and until an adequate theoretical basis is developed it is likely that this applied economic analysis of policy will be conducted in an ad hoc way in the process of empirical research.

Managerialism and Weberian political economy

Max Weber's social theory of action, which underlies the managerialist approach, also offers a possible theoretical foundation for the 'applied economic analysis of policy' approach proposed by MacLennan. Indeed, it has been claimed that 'Weber's real achievement lay... in achieving a substantive reconciliation of economics and sociology so as to make possible a unified, though differentiated, liberal social theory' (Clarke 1982, 213). If this is the case, then Weberian social theory is of interest not only because of its underpinning of managerialist housing research but also because of its potential contribution to an applied economic analysis of policy approach.

Marginalist economic theory was accepted by Weber as a rationalist explanation of the working of the competitive market economy. However, he emphasised the importance of considering not only an individual's rational orientation to self-interest (utility maximisation) but also the role of organisations in society.

> Where the economic institutions of the market – money, etc. – provided the necessary formal framework for rational social action, the social institutions of organisations...provided its contingent historical framework. It was on this basis that Weber differentiated between...'economics' and 'sociology'.... Weber's sociology was primarily concerned with establishing a typology of organisations according to the ends which motivate their formation and inform their direction; the means available to those ends, the value-orientation of action typical to them and their internal dynamics. It was quite possible and indeed very likely that the formation of these organisations would subvert the formal rationality of the competitive market system, for they were established precisely to achieve ends that could not be achieved directly through rational economic action.
>
> (Clarke 1982, 216)

Clarke argues that, by placing the individual's rational voluntaristic action at the centre of his social theory of action, 'Weber developed the conceptual foundations for both modern economics and modern sociology. These foundations were then classically elaborated for modern economics by Lionel Robbins in *An Essay on the Nature and*

Significance of Economic Science and for modern sociology by Talcott Parsons in *The Structure of Social Action* (Clarke 1982, 234). The distinction was that economists abstracted economic relations from all social content by according primacy to economic rationality as an ethical ideal, whereas sociologists viewed political, religious, moral or aesthetic criteria as equally valid orientations for social action. The areas of enquiry indicated by MacLennan as necessary to the applied economic analysis of policy appear to imply a retreat from the primacy of economic rationality, and in so doing raise the possibility of developing a Weberian political economy in which other forms of rationality are admitted, while still focusing on the individual and his subjective motivations. The question of whether or not Weberian social theory can be employed to inform the broadening of economists' concerns is returned to later in this chapter.

Weberian social theory already underlies one major strand of housing research – the managerialist approach. This relies in particular on Weber's concepts of class and status, and it will therefore be helpful to explain these terms before reviewing their use in housing research.

As we have seen, 'the basic concept in Weber's sociology is that of the human subject endowed with free will who, in interaction with others, attempts to realise certain values or objectives' (Saunders 1981, 25). For Weber, class was an analytical category in which such individuals were grouped together on the basis of their common economic situation in relation to commodity and labour markets. 'Although he followed Marx in arguing that classes were objectively-defined categories with a material base, Weber differed from Marx in arguing that classes could arise in any market situation' (Saunders 1980, 67) and not only in relation to the means of production. Thus, he used the term 'class' to refer to all individuals with the same chance of procuring goods, gaining a position in life, and finding inner satisfaction, 'a probability which derives from their relative control over goods and skills and from their income-producing uses within a given economic order' (Weber 1968, 302). In this way Weber 'located class analysis in the sphere of distribution rather than production relations, in the market rather than in the mode of production' (Saunders 1981, 138).

Since class denotes individuals sharing a common economic interest that derives from a common economic situation, classes may be seen to arise objectively out of differential market situations, and do not require to be subjectively recognised either by members or outsiders. Class is not therefore the sole basis of social action for, as Clarke (1982, 218) points out, 'there is no reason why individuals should necessarily be aware of their common class situation and still less why they should necessarily establish class organisations on that basis'. While conflicts of economic interest are inevitable in capitalist societies, so that class formation is inherent, classes may or may not play a significant role in action and organisation. Instead, Weber's pluralistic conception highlighted other forms of organisation which he termed status groups and parties. Status groups, unlike classes, are subjectively recognised by their members since they are defined in terms of similarity of lifestyle, reflecting the distribution of social honour or prestige in society. While prestige may be closely related in practice to patterns of economic inequality, Weber argued that it is nevertheless analytically distinct and that status 'need not necessarily be linked with a class situation' (Weber 1968, 932). Thus, 'classes are stratified according to their relations to the production and acquisition of goods; whereas status groups are stratified according to their principles of their consumption of goods as represented by special styles of life' (Weber 1968, 938).

Rex and Moore (1967) drew upon the Weberian concept of class in their seminal study of competition for housing in Sparkbrook, Birmingham. They distinguished housing classes according to households' differing degrees of access to private and public housing. Two key criteria were identified: the size and security of income were crucial in gaining mortgage-funding and so gaining access to owner-occupation, and the ability to meet the need and length of residence qualifications applied by the local authority were crucial in determining access to council housing. Groups such as low-income immigrants who could satisfy neither of these criteria were therefore disadvantaged both in the competition for the most desirable type of property (assumed to be suburban owner-occupied housing) and in the competition for the 'new public suburbia' (council housing) created by the collective political action of the less affluent (Rex and Moore 1967, 9) and they were therefore obliged to seek accommodation in the declining and decaying private rented sector. On the basis of these structures of allocation, Rex and Moore identified six housing classes:

1. The outright owner of a whole house
2. The owner of a mortgaged whole house
3. The council tenant
4. The tenant of a private landlord
5. The home-owner who sub-lets to repay a loan
6. The tenant of rooms in a lodging house

The study was influential in three respects: it related housing research back to Weberian theory, it focused on differential access and its relationship to the individual's economic situation, and it shifted the emphasis of research towards the analysis of institutions.

Although the work was highly influential, it was also widely criticised. For one thing, Rex and Moore's classes were static outcomes of housing allocation processes, rather than an account of the dynamics of household mobility. Haddon (1970) argued that Rex and Moore had confused a model of potential power in the housing market with a typology of households based on current housing tenure: the problem is that people who presently share the same housing tenure need not also share a common capacity to gain access to a more favourable type. For example, some may be trapped in private renting, while others may be tenants purely by choice. It follows that 'the emphasis of the analysis ought to be on the means and criteria of access to desirable housing, and the ability of different people to negotiate the rules of eligibility' (Haddon 1970, 129). This, however, is not a criticism of housing class models *per se*, but merely of Rex and Moore's empirical application.

Another difficulty concerns the assumption that all households seek the same type of house (suburban home ownership). This was found by later research to be unjustified (Davies and Taylor 1970; Couper and Brindley 1975) and consequently the basis of class conflict is unclear. In any case, Rex and Moore appear to have erred in relying on preferences at all, given the objective basis of class. As Saunders (1980, 73) explains:

> Both Marx and Weber argued that classes are identified objectively, and there is no reason to suppose that those sharing a common class situation should also share a common recognition of their situation, or a common set of values about it. The problem as far as Rex and Moore's work is concerned is that, although they argued that housing classes did not necessarily share a common class consciousness, they nevertheless situated their analysis in the process of competition (which necessarily takes place on the basis of subjective preferences) rather than conflict (which can be identified in terms of conflicts between objective class interests, and which need not therefore depend upon subjective preferences).

Most damagingly, from a theoretical point of view, Haddon (1970) argues that what Rex and Moore have identified as housing classes are in fact housing status groups, within which people share similar lifestyles rather than similar market situations. Haddon suggests that a distinction must be drawn between the housing market (in which the structures of allocation and access give rise to different patterns of consumption, and hence to differences of housing status), and the domestic property market (which gives rise to genuine class divisions based on the capacity to realise financial returns and/or income from the sale or rent of property). This criticism has led Saunders (1980) to argue that tenure must be seen as a key element in housing class situations, since landlords and especially owner-occupiers own housing not merely for consumption but also for its accumulative potential, whereas tenants do not. In Saunders' view this implies the identification of three housing classes, each of which may be further subdivided: these are the owners of private capital involved in the supply of housing and its associated services, a middle class of home owners for whom housing is both a consumer good and an investment, and non-owners of domestic property (i.e. tenants and homeless people). The elaboration of these classes in a rural context is attempted in Chapter 4.

These criticisms have led to various developments in housing research, of which perhaps the most obvious has been an increased concern with access to housing, focusing particularly on the individuals within organisations and bureaucracies who control access to housing. These would include public housing officials, planners, local government officers and councillors, property developers, estate agents, building society managers, social workers, landlords and many others. The managerialist perspective thus initiated was elaborated by Pahl in a number of papers (collected in Pahl 1975). According to Pahl, (1977, 50) managers

> exert an independent influence on the allocation of scarce resources and facilities which may reinforce, reflect or reduce the inequalities engendered by the differentially rewarded occupation structure.

It has generally been accepted that in his original formulation, Pahl 'under-emphasised the extent to which market forces and the activities of central government seriously constrained the managers' actions', but that in a later reformulation (Pahl 1978), he 'continued to stress the managers' role in resource allocation, but recognised that they had little control over resource availability' (Goldsmith 1980, 42). He thus acknowledged the constraints imposed on managers by wider political and market factors, while continuing to stress the pivotal role of managers and gatekeepers in resource allocation. However, Morcombe (1984) disputes this accepted reading, arguing that Pahl's original work was misinterpreted and distorted and that it is this misinterpretation which has been responsible for overemphasising the independence of urban managers and neglecting the wider socio-economic structure.

A large number of empirical studies were devoted to analysing the role and effects of various urban managers, following Pahl's ideas, and these established that managers in many spheres of housing supply exert a major influence over distributional outcomes. For example, Niner (1975) examined the role of public sector housing managers in council house allocation in the West Midlands; Gray (1976) found that officials' assessments of tenants in Hull had a significant influence on the quality of property offered to them; Karn (1976) documented the selectivity of local authority lending in Birmingham; and the 'redlining' policies of building societies have been revealed by a number of studies (Boddy 1976; Lambert 1976; Weir 1976; Williams 1978). Other managerial actors studied include private landlords (Elliott and McCrone 1975; Short 1979), estate agents (Palmer 1955; Burney 1967; Williams 1976), developers (Craven 1975; Ambrose and Colenutt 1975), planners (Dennis 1972; Buchanan 1982) and landowners (Robson 1975).

> However, their concern was wholly with the managers themselves rather than seeing them, as did Pahl, as a means to an end. They did not accept all Pahl's notions though a number of over-generalisations have recently taken place which have both termed them as urban managerialism and attributed them to Pahl (Duncan 1977). The result has been a distortion of Pahl's original notions.
> (Morcombe 1984, 13)

Morcombe argues that Pahl always understood that managers acted within the constraints of the socio-economic system, but that he proposed to observe the workings of the socio-economic system (his true object of study) through the window of the actions of urban managers: 'an examination of the managers' actions would not only illustrate the constraints imposed by a socio-ecological system upon an urban population but would also identify those constraints' (Morcombe 1984, 13). Whether or not one accepts Morcombe's re-interpretation, Pahl subsequently introduced two important refinements to his later work on urban managerialism which overcome the main weaknesses of his earlier proposals (Saunders 1981, 121–36).

First, Pahl distinguished between managers in the private sector and those in the local state sector, focusing his analysis on the latter. Second, he revised (or clarified) his view of the autonomy of such local officials by explicitly recognising that they had to operate under constraints imposed both by their relations with the private sector and with central government. Urban managers were therefore seen not as independent agents, but as middlemen performing 'crucial mediating roles both between the state and the private sector and between central state authority and the local population. Another set of private managers control access to capital and other resources' (Pahl 1977, 55).

It follows that analysis of the mediating role of officials at the local level must be complemented by an analysis of the autonomous actions of the central state and of the private sector's pursuit of profitability, and this led Pahl to develop a theory of corporatism to address these issues. It is not necessary for the purpose of this book to include a detailed account of corporatism, since the scope of this book is mostly restricted to the operation of constraints and their consequences, leaving analysis of the origins of these constraints as a task for future research. It is sufficient to point out the difficulty inherent in the empirical application of the corporatist approach, because

> Pahl's recent work leads empirical research into the familiar problem of the receding locus of power; the actions of urban managers can be understood only in the context of the crisis of national state policy; national state policy can be understood only in the context of the operation of a complex mixed economy; the operation of the economy can be understood only in the context of the crisis of the capitalist world; and so on. Thus the researcher who starts out by attempting to understand, say, the patterns of housing inequalities in Birmingham ends up by trying to analyse the oil policies of the Middle Eastern states or the impact of American fiscal policy on the international balance of trade.
>
> (Saunders 1981, 135)

It is for exactly this reason that an attempt has been made to set bounds on the level of analysis of this book, and it is worth noting that Weber supported such a partial approach, arguing that total explanations were impossible, and that research must proceed by selecting partial aspects of the social world for study on the basis of ideal type constructions.

Marxist political economy

Much recent urban housing research has been undertaken from a Marxist political economy perspective, which emphasises the role of the state in housing markets in support of the interests of capital. Marxist theory asserts that fundamental class divisions relate to production alone, with class conflict arising between the owners of the means of production and those who sell their labour power: housing tenure divisions are not the basis for class divisions, but are merely ideological, diverting attention from class conflicts and encouraging a fragmentation of the working class. The role of the state is to support capital accumulation, by ensuring the continuing dominance of the working class by the capitalist class, while at the same time ensuring the reproduction of labour power.

The two most influential schools of Marxist urban research are those represented by Harvey and Castells, each of whose work has also been widely criticised. Harvey (1973,1974,1977) has attempted to develop explanatory links between capital accumulation, the role of the state and urbanisation (MacDowell 1982). His approach emphasised the interlocking relationships between the financial institutions and the state, 'whose structuring and restructuring of the residential environment functioned to mould consumption patterns and fragment class relations and contributed to the social stability of the system' (Morcombe 1984,

20). Further, he viewed the growth of owner-occupation as a response to the crisis of capital over-accumulation, helping generate demand for household goods to absorb the structural surplus of capitalist production. Saunders (1981, 228–31) has pointed to two weaknesses in Harvey's work: first, his work 'fails to address the central question [to a Marxist] of how investment in urban infrastructure may affect the rate of profit in the economy as a whole'. Second Harvey's work only acknowledges class struggle [again central to Marxists] as a secondary factor, 'since the major contradiction of over-accumulation is not brought about by the working class at all, but is simply the result of competition between individual capitalists'. MacDowell (1982, 84) points also to ambiguities in Harvey's analysis of the role of the state, since it is unclear whether the state is a mere adjunct of financial capital, or whether it mediates between industrial capital and financial capital, or mediates between capital and labour.

Castells (1977) also highlights the state's role in regulating the economy to maintain capital accumulation, but his approach does not rely upon the theory of capital over-accumulation. Instead he argues that the state must intervene to provide social and economic infrastructure for collective consumption both in order to maintain overall profit levels and to ensure the reproduction of labour power with minimum class conflict. Housing is such a form of infrastructure and therefore 'stands in a dual relationship to the production process. It must not only be produced for profits to be accumulated, but also consumed by the workers who are an essential part of the production process.' (MacDowell 1982, 84). The need for direct state provision results partly from the inherent contradiction between the interest of employers in keeping housing costs (and therefore wage demands) down, and the interest of housing producers in raising house prices. Apart from the conflict between these two fractions of industrial capital, the state has to intervene to ensure the health and compliance of the workforce. Castells argues that there is inherent potential for a crisis in the provision of commodities necessary for the reproduction of labour power because capitalist production

> is concerned with exchange values while consumption is concerned with use values. There is, in other words, no necessary reason why what it is most profitable to produce should coincide with what is most socially necessary to consume, since the investment of capital is dictated by rates of return rather than need.
>
> (Saunders 1981, 188)

'If this growing contradiction, which becomes manifest in housing shortages, inadequate medical care, lack of social facilities and so on, is not regulated in some way, then it must necessarily create new forms of political tension and strife,' (Saunders 1981, 189), and this necessitates state intervention.

Once again, Saunders (1981, 189–218) has provided a comprehensive critique, concluding that 'the fruitful residue from his [Castells'] writings represents only a small part (and a considerably amended part at that) of his overall perspective'. Amongst the many serious weaknesses which he identifies in Castells' work, the most critical is that 'Castells' theory may be rejected on grounds "internal" to its own discourse (namely that it is premised on a theory of action, which it denies, and that it requires counterfactual conditions, which it cannot identify)' (1981, 218). The concept of collective consumption itself has been the subject of much criticism, for example, in terms of its unnecessary spatial delimitation, and its incoherent specification (1981, 211, 216; Dunleavy 1979). Despite these defects and the inadequacy of Castells' theories, both Saunders and Dunleavy regard a modified concept of collective consumption as a useful ideal typical construct, and it is possible that this may have some potential application in rural housing research.

Ultimately, though, a methodological basis for research must be chosen for its apparent usefulness and relevance, and its internal consistency. While the concept of collective consumption may hold out some promise, it rests on less certain foundations, and appears less potentially fruitful, than a Weberian political economy approach, as outlined on pp 28–31. Because of the weaknesses of Harvey's and Castells' approaches, a Marxist political economy approach is not explored in this book. Nevertheless, the pursuit of a Weberian approach should not rule out the possibility of important insights being gained in future rural housing research which follows the 'urban sociology' approach outlined in the final chapter of Saunders (1981) and elaborated in Saunders (1986), developing both Pahl's managerial focus and Castells' concept of collective consumption.

A methodological basis for rural housing research

The possibility was raised earlier of developing a Weberian political economy approach, drawing both from marginalist economics and from Weber's social theory of action. It has been shown that these two elements are compatible once the non-primacy of economic rationality is conceded. Such an approach, it is argued, would allow economists to broaden the scope of their analysis to admit social and political concerns – often relegated by agricultural economists to the status of the error term in a regression equation according to Newby (1982). While a minority of agricultural economists have pursued a 'tradition of policy analysis' which admits social and political factors (Newby 1982, 136), this work tends to proceed in an ad hoc fashion without any explicit theoretical foundation for the broadening of the analysis to embrace social and political factors.

A neo-classical economic analysis of policy would be concerned with issues of economic efficiency in the production, allocation and consumption of goods, and with issues of equity between consumers in their distribution. When applied to rural housing, such an approach would lead to a focus on the efficiency of rural housing markets, the possible existence of public good effects or externalities, the relationship between house prices and incomes, and an assessment of distributional outcomes. Policies would be assessed in terms of their effects in distorting or correcting the market, and in terms of losers and gainers from policy's implementation. The twin themes of this book – the competition for land between landscape preservation and housing provision, and differential access to housing – can each be seen to derive from this conventional economic approach as issues of efficiency and equity respectively.

A Weberian approach would indicate different, but related, analytic concerns. These would include the identification of housing classes and, perhaps, housing status groups as a basis for the conceptualisation of conflict within rural society. While this would also (as with the economists' approach to equity) suggest an investigation of differential access to housing, it is clear that Weberian theory implies a concern not merely with access to housing as a consumer good but more fundamentally with access to domestic property ownership as a means to capital accumulation. In the context of the conflict of interest over rural housebuilding between home owners already living in rural areas who wish to preserve their amenity and hence their property values and those denied this means to capital accumulation, this aspect of the Weberian approach also highlights the competition for land between landscape preservation and housing provision as a central issue. The two themes of this book are therefore suggested both by Weberian concerns and by a conventional applied economic approach.

At the same time, Pahl's revised formulation of the role of urban managers as local state officials mediating between central government and local interests, and between private profit-seekers and social needs, offers a further useful analytic conception which may be applied to the role of institutions intervening in rural housing markets. It has been accepted that managers and organisations operating within the rural context are themselves subject to constraints which limit their freedom of action and their ability to make resources available for housing. As will be seen, these constraints play a major part in determining distributional outcomes, and can hardly be ignored. Such constraints derive both from central government's control over local decision-making and from other sources, such as the budgetary crisis of the CAP and the technological changes in agriculture and forestry which have reduced these industries' demands for labour. It is important therefore to consider the nature of such constraints, and also to analyse them in terms of their distributional consequences for housing classes.

Weberian concerns may therefore be admitted by economists as refinements to their own ad hoc approach to social and political concerns rather than as contradictory alternatives. On the simplest level, an analysis of the distributional consequences of policy or of differential access to housing presupposes some form of typology of those affected. Even elementary economic models distinguish between producers and consumers, firms and households, taxpayers and government. In the rural housing context, it would seem appropriate for economists to draw on Weber's social theory of action in constructing such a typology and therefore to try and identify housing classes, and perhaps housing status groups, in the Weberian sense. For example, a distinction between those who can buy into owner-occupation and those who must rent is fundamental to any policy analysis. The basis of such a typology is discussed further in Chapter 4. At a more fundamental level, it has been argued that Weberian social theory offers more than a device for assessing distributional outcomes, since it also helps to conceptualise the nature of social and political forces determining access to rural housing, and moreover suggests research questions which may be pursued by economists and sociologists alike.

This book will therefore proceed by addressing the two themes already identified in Chapter 1. The conflict of objectives inherent in policy towards rural housing in Britain – landscape preservation versus housing provision – may be viewed by the economist as an instance of market failure, requiring corrective government intervention, and from a Weberian perspective as the manifestation of fundamental class cleavages between housing classes distinguished according to their differential market positions in relation to potential capital accumulation from the ownership of rural housing. The economist's concern for equity in the distributional outcome of the rural housing market clearly relates to this Weberian class analysis, as well as to the focus on access to housing (for consumption) deriving from Rex and Moore's work. The analysis pursued here stops short of attempting to identify the ultimate locus of power, and focuses instead on intervention at a local level and on the constraints which central government imposes on local action. This, as we have seen, is consistent with Weber's insistence that only a partial analysis is possible, and follows the familiar managerialist approach in that 'although the role of intermediary governmental agencies (usually those which are locally based such as local housing authorities) is recognised, their behaviour is studied in an exogenously determined environment in which the overall structure of political and economic power is taken for granted' (MacDowell 1982, 82). This is not intended to preclude, or diminish the worth of, future studies of more fundamental political and economic structures as they relate to rural housing, but merely to exclude them from the explicit concerns of this book. The question of the partial scope of this treatment is returned to again in the final chapter.

Chapter three

Rural planning and housing policy – a conflict of objectives

Public intervention presupposes policy objectives both at central and local government levels. According to Rogers (1983,124) policies for rural housing in advanced Western societies are generally designed to fulfil one or more of three broad objectives: a) as an adjunct to agricultural policy; b) as a means towards wider rural development and helping the poor; c) as a vehicle for landscape protection and conservation. In Britain, Rogers suggests, the third objective has been particularly characteristic, placing the emphasis on the control of housing in the countryside rather than on its provision.

From the point of view of the economist, each of these objectives may be resolved into the pursuit of global objectives of efficiency and equity, but policy-makers will have more specific objectives, whether implicit, explicit, or imposed by statute. This chapter begins by considering, from an economic perspective, the statutory objectives of countryside policy and their relevance to rural housing. Subsequently, the statutory basis of housing policy is also considered, and their interaction is discussed.[1] This leads to a discussion of the central post-war policy objective, that of containing settlements and preventing housebuilding in the countryside. This is crucial to the analysis because of the conflict of interest over land release for rural housebuilding which exists between home owners in rural areas and those denied this means to capital accumulation, an issue which forms the basis of Weberian class divisions (Saunders 1980).

Objectives of countryside policy – the economist's view

The Countryside Act 1968, section 11 and the Countryside (Scotland) Act 1967 place a duty on every minister, government department and public body, in exercising their functions in relation to land, 'to have regard to the desirability of conserving the natural beauty and amenity of the countryside'. Local authorities also have powers, under numerous acts (and national park authorities and the Countryside Commission have a duty under the 1968 Act) to promote public enjoyment in the countryside. A further duty 'to have due regard to the interests of agriculture and forestry and to the economic and social interests of rural areas' is imposed on every minister, the Countryside Commission, the Nature Conservancy Council and local authorities by section 37 of the 1968 Act. Public policy, then, involves the preservation of landscape beauty, the promotion of public enjoyment, and the maintenance of rural communities.

In theoretical terms, the pursuit of each of these objectives may be viewed from the point of view of the economist as justifiable public intervention in the land market, on the ground that each promotes a land use which would be undervalued in the free market. According to Price (1978, 16), 'landscape is a classic public good: it is difficult for its owners to exclude the public from consumption of it; and the product is not used up in the process of consumption.' Landscape preservation and public enjoyment will frequently be jointly produced, and Price's term 'landscape' seems to imply some such joint product. Yet it will be useful initially to distinguish between landscape preservation and the recreational enjoyment of the landscape, however inseparable these may appear in practice.

Landscape preservation in this narrower sense is a pure public good. Regardless of whether an individual visits or wishes to visit the countryside, he is a consumer of landscape preservation in the same way that he is a consumer of defence, and his consumption is not reduced by others' consumption. Much of the rhetoric of conservation is couched in terms of the landscape's utility for generations as yet unborn. Price accepts the argument that there is a 'value placed on simple knowledge of a scene's existence, independently of it being experienced by anyone', although he regards this argument as 'too metaphysical' for him to pursue (1978, 27). Nevertheless, this same argument appears to be implicit in the government's acceptance of the Sandford Committee's recommendation that landscape preservation should be given precedence over public enjoyment of national parks if these aims are irreconcilable (DoE 1976).

There are clear implications for policy (Willis 1980, 54–7). The socially optimal price to the consumer of such a good is zero, because the marginal cost of adding an extra user is zero. Yet the optimal price to the producer of landscape preservation is positive. Obviously, no price can simultaneously be both zero and non-zero, and the price system can be seen to be inherently incapable of dealing with such cases (Baumol and Oates 1975). To meet this double requirement a subsidy to the producers of landscape preservation is necessary, financed from general taxation (Willis 1980, 56–7), as paid to farmers in ESA, for example.

The second output, the public enjoyment of the countryside, is more diverse in form. Visits can range from formal, site-based recreation, such as water-skiing or visiting historic houses, to informal, extensive activities such as hill-walking, picnicking and general sightseeing. Instances of the former type are not public goods, in that exclusion is feasible and congestion often occurs. If an activity such as water-skiing creates negative externalities, the remedy is clear: a tax may be imposed on consumers or the activity may be regulated. It is the informal, extensive activities which may be seen to have some of the characteristics of public goods, and which therefore pose greater problems in policy formulation.

Hill-walking, for example, might in principle be priced, but in practice it is rarely worthwhile to try and charge consumers, or to exclude them, because the transactions costs would exceed the revenue. Such forms of recreation may therefore be regarded as quasi-public goods on the basis of their non-excludability of consumption, even if there is some degree of rivalry in consumption as a result of visitor pressure.

The problem here for the policy-maker is not one of poorly defined property rights, but one of finding an efficient way of charging the consumers of informal recreation. Most direct methods will incur prohibitive transactions costs, and indirect methods are likely to introduce other distortions. For example, visitors might be charged when they park their cars to walk or to picnic, but this would have the (perhaps beneficial) side effect of discouraging private cars. Again, a tax on visitors' accommodation might be imposed but this would have the (undesirable) side effect of decreasing local incomes. Most attempts at charging will require government to intervene in their implementation, since they are outside the individual landowner's jurisdiction.

If there is no efficient means of charging consumers, either because the transactions costs are prohibitive or because the mechanism would introduce unwanted side effects, then it may be justifiable for government to pay the landowner or farmer to 'produce' public enjoyment of the national parks, as with access agreements and the Upland Management Scheme. In these cases it is the general taxpayer rather than the user who is charged.

In reality, landscape preservation and the public enjoyment of national parks may often be joint products. Where they are jointly consumed, the above discussion suggests that the marginal cost to the consumer should be zero until visitor pressure on the landscape or congestion begins to impose external costs. Beyond that point, some form of management should be practised which will deter additional visitors while continuing to protect the landscape. This is implicit in government advice that the recreational use of national parks should be related to their particular qualities and capacities (DoE 1976), for example, in the Lake District Special Planning Board's policy of preserving 'the quiet character of extensive areas of the national park, areas which should be left to those who deliberately set out to seek them... where an impression of remoteness should be maintained' (LDSPB 1985, 136–7).

As joint products in production, neither landscape preservation nor public enjoyment yields a price to the producer which will induce him to produce a socially optimal output. In each case the policy conclusion is that he should be paid from general taxation to produce at the required level.

The use of land to sustain local communities was not initially seen as a legitimate policy objective.

> Although the Scott Committee had affirmed that healthy rural communities were essential to the preservation of the countryside, there had been no recognition of this relationship in the 1949 Act; attention being drawn only to the needs of agriculture and forestry.
>
> (TRRU 1981, 26)

The welfare of wider rural communities *per se* proved to be of little concern to policy-makers at this stage and, for example, 'it was not until the late 1960s that any of the north-west's planning authorities began to understand the need for active policies to protect the life of upland communities' (Capstick 1980, 59).

When the socio-economic problems of rural communities were publicised in the 1970s, they were presented as 'rural deprivation', a vague welfare concept usually attached to numerical evidence of low levels of service provision (the 'arithmetic of woe') and concern about rural depopulation (McLaughlin 1980). The arguments for favouring the economic and social interests of rural communities were therefore arguments of social equity. Council housing, buses, shops and even employment were seen as objects of distributive inequity, to which people in rural areas had an equal claim.

However, it is now being realised that the welfare of rural communities is not only a matter of equity but also of efficiency. The public good aspects of the countryside, landscape preservation and public enjoyment, depend upon the survival of local communities. The national parks, for example, are living, man-made landscapes.

> The emphasis on appearance has been a handicap both to understanding and to action. It has encouraged the idea that problems deeply rooted in the exploitation of the resources of the national parks for economic purposes can be dealt with by protective

> designations of limited areas, while ignoring the social and economic trends or policies that are literally changing the face of nature.
>
> (MacEwen and MacEwen 1982, 111)

Thus landscape cannot be preserved, except by implementing social and economic policies which sustain working communities and encourage appropriate land management. The public enjoyment of the countryside is also diminished in the absence of a local population, not only as 'props on a rustic stage' (Pahl 1966) but also as suppliers of accommodation, food and other services. 'This means that promoting the health of local communities...will contribute to the value of these areas as places of recreation for the urban majority, provided the methods which are used do not destroy their rural character' (HM Treasury 1976, 43).

The argument being proposed is essentially that the economic and social welfare of rural communities is a form of producer good, essential to the production of two public goods, landscape preservation and public enjoyment. Resource allocation for the economic and social benefit of local communities may thus be encouraged not only from an equity perspective but also on efficiency grounds. This conclusion has particular relevance for rural housing policy in areas like the Lake District, where it has been argued that the survival of rural communities is threatened because 'there is no housing available in their native valley for the young people on whom the future of the community depends' (Feist *et al.* 1976, 95). While the control of new housing development may be justified on the grounds of landscape protection, the provision of housing which helps to sustain the rural community is equally necessary to the long-term objectives of countryside policy.

Objectives of housing policy: the economist's view

Housing policy under most post-war British governments has also been concerned with correcting the free market's inefficient allocation of resources. 'Over the years the view has been held that the well-being of present and future generations is better safeguarded if housing is not left solely to market forces' (HMSO 1977, section 2.15).

Charles (1978) lists several of the imperfections of a free housing market. These include price instability, lack of information, the durability, immobility and heterogeneity of housing, and the lengthy construction process. In contrast to Lansley (1979) and Balchin (1985), she argues that these shortcomings do not in themselves justify public intervention, because the complexity of housing markets is not a severe source of market failure; and while housing market imperfections are clearly present, government interference itself is rarely perfect.

Externalities, in particular, 'are highly localised, affecting only a few households, and they are not particularly disruptive': therefore, 'there are no externality grounds for wholesale market interference of any type', although 'certain policies aimed at correcting particularly gross external effects' may be justified, for example, to eliminate urban slums. This would allow the argument that the public good aspects of rural housing discussed, might provide the justification for similar specific policy measures. Further, as MacLennan (1982a, 145) points out, if the government intervenes in other areas of policy, such as through landscape protection, 'then complementary or compensating distortions may be required in the housing system'.

Generally, however, it is less on grounds of inefficiency than on grounds of inequity that intervention in housing markets has been justified. McDowell (1982, 92) points to the fundamental weakness of the Paretian concept of efficiency under which the 'analysis of consumer preferences and satisfaction is based on the continued existence of a wide degree of inequality': whether such an outcome is acceptable or not is a value judgement, and both Nath (1973) and Rowley and Peacock (1975) have clearly demonstrated the inherent conservatism of the Paretian ethical assumption.

Lansley (1979, 21) outlines the case for government intervention on grounds of equity:

> Even if intervention were to correct for such imperfections in order to promote an efficient allocation, the market would still produce an unacceptable distribution of housing resources. The principle of equity embodied in the market view of the working of the economy – that output is distributed in accordance with consumers' own preferences – has two major weaknesses. In the first place, the distribution of market-expressed preferences depends upon the existing distribution of income. A free market would be unable to provide housing of a socially acceptable level to those with low incomes.... Secondly, even if the distribution of income could be made less unequal, there are other views of equity than that contained in the consumer-sovereignty outlook.

The concept of a socially acceptable level of housing is discussed in more detail in Chapter 4, where it is noted that successive governments have espoused the 'traditional aim of a decent home for all families at a price within their means' (HMSO 1971, 1; HMSO 1977, 7). McDowell (1982, 54) notes that government intervention in housing markets is common to all advanced industrial nations and that while the form of intervention varies the aims are similar: 'The first is to maintain minimum physical standards... The second aim is to assist particular sectors of the population to acquire housing' either through direct public provision or through subsidies. To justify such intervention the slum housing conditions of nineteenth-century London and other cities are cited as evidence of the inequity of the free market (Hamnett 1976, 39).

The outcome, as Lansley (1979, 39) argues, is that housing has been regarded in Britain as something between an ordinary consumer good and a social service, and thus housing provision 'has been characterised by a mixture of unfettered free enterprise, free enterprise subject to controls and public enterprise, with slight shifts of emphasis at different times'. MacLennan (1982a, 167) argues that British housing policy 'lacks a central organising principle' so that while housing market intervention may be rational and required, the present distribution of subsidy for example 'is an outcome of chance, history... and tenure selected'. 'The broad goals of policy, let alone the intentions of specific instruments, are seldom politically defined in operational terms' (MacLennan 1982a, 163), and so it becomes difficult to evaluate policy outcomes against objectives.

Since 1979, however, the objectives implicit in the government's housing policies have been fairly clear. The overriding imperative has been the requirement to cut the public expenditure devoted to housing, as a consequence of the monetarist macroeconomic policy pursued. This necessity coincided with two ideologically derived objectives, namely the reduction in size of the public sector in housing and the increase of council house rents to unsubsidised levels (Gillett 1983, 54). The promotion of home ownership and a 'property-owning democracy' have been the achievements claimed by government ministers, but this contrasts with MacLennan's (1982b) characterisation of Scottish housing policy in the 1980s as 'public cuts and private sector slump'. He argues that:

> policy has almost entirely consisted of public expenditure cuts. These reductions have not been accompanied by policy restructuring to enhance efficiency and distributional effects and, in particular, the rate of capital spending has been neglected. Insofar as privatisation has proceeded it has depended upon lump sum wealth transfers, via council house sales, rather than upon a flourishing private housing sector.
>
> (MacLennan 1982b)

The objectives underlying these policies thus appear to be macroeconomic rather than housing objectives, relating to the Public Sector Borrowing Requirement rather than to housing consumption and production.

The 1988 Housing Act was an attempt to correct these weaknesses, addressing the rented sector as well as home ownership. The objectives of this legislation were elaborated in the 1987 White Paper *Housing:*

The Government's Proposals. These were: to encourage the further spread of home ownership; to revitalise the private rented sector; to achieve greater efficiency in the use of public money and to increase the role of the private sector; and to alter the role of local authorities from providers to enablers. Each objective relates to efficiency rather than to equity, emphasising the government's aim of reducing public intervention in housing, except in relation to subsidies for home ownership.

Rural housing policy – a political economy perspective

Although the responsibility for rural policy is divided between a number of ministries, planning, conservation, recreation and housing functions, all fall within the remit of one ministry – the Department of the Environment (DoE) in England and the Scottish Development Department (SDD) in Scotland. The reconciliation of countryside policy objectives and housing objectives at national level is therefore very largely a matter for the Secretary of State for the Environment and his Scottish equivalent, the Minister for Home Affairs and the Environment. Each has made statements to the effect that it should be possible to combine strict controls on development in the countryside with housing provision for the local population.

Thus, the Labour government in 1976 responded to the Sandford Report on national parks as follows (DoE Circular 4/76):

> The national parks are areas whose amenity has to be conserved, but it should be possible to meet the legitimate housing needs of those who live there by providing residential development in suitable locations.

The Conservative government encountered the same conflict in relation to the Lake District: the Secretary of State (DoE 1983) recognised that because of the need to protect the area's outstanding landscape

> the provision of housing within the Lake District National Park presents particular difficulties and he has sympathies with the Planning Board's wish to ensure that sufficient housing is made available for people who live and work in the area.

This government's attempt in 1983 to alter the balance between countryside protection and land release for housebuilding in green belts foundered on considerable opposition from within its own party (Nuffield Foundation 1986, 26). In Scotland, on the other hand, the balance between these objectives in the remoter areas changed when restrictive policies, applied since 1960, were amended by SDD Circular 24/85 to allow local planning authorities to relax restrictions on development in the countryside by defining in their structure plans :

> the general circumstances and criteria for areas within which well designed and well located development, which does not affect land that is important for the maintenance of agriculture, damage the scenic or nature conservation interest of the land, or make undue demands on public services, can be successfully merged into open countryside.

Despite this recognition of the potential (and actual) conflict between rural housing policy objectives, there are relatively few policy measures specific to rural housing in Britain. Apart from a recent increase in funding for NACRT, the policies which attempt to resolve this conflict address only the competing claims for land, through land-use planning. Yet this inherent conflict between housing provision, for reasons of both efficiency and equity, and landscape protection, on grounds of efficiency, is a fundamental distinguishing characteristic of rural housing policy, since it may also be regarded as the basis of class conflict between owner-occupiers and non-owners whose market situations in relation to land release for housebuilding are directly opposed. This is discussed in detail in Chapter 4.

It is therefore relevant to summarise, at this point, the generally restrictive rural settlement and development control policies which have been pursued by both central and local government since the war, and then to discuss the relatively few specific provisions which have been made to address the conflict between these policies and housing objectives. These provisions have attempted to take account of the special housing needs of farm and forestry workers, to restrict resales of council houses to local people in areas of high second home numbers, and to cater for local needs.

Containing settlements

The issue of rural housing received some prominence in the late 1930s and early 1940s. In 1936 the Minister of Health established a rural housing sub-committee of the Central Housing Advisory Committee, and this issued three reports in 1936, 1938 and 1944. The last of these reports proposed a framework for specifically addressing rural housing problems through joint committees for each county, but although the committees were established their work lost impetus as central government's enthusiasm waned (Rogers 1976, 95–6). Since that time, housing policy has not generally distinguished between rural and urban areas, and the rural policy most relevant to rural housing provision has been rural settlement planning. As Newby (1980, 185–6) argues,

> The subsuming of rural housing under a general 'housing problem' has also been accompanied by a change of emphasis in

> housing policy. Between the wars the aim of housing legislation was simply to stimulate the construction of as many houses as possible in rural areas; since 1945 the aim has been to control the number of houses in rural areas as part of overall planning policies designed to contain the growth of urban sprawl, prevent the loss of good agricultural land and protect the visual quality of the countryside. Far from encouraging local authorities to build more rural houses as in the 1930s, there has been an active discouragement, involving the imposition of strict planning controls, particularly over housing in the open countryside and in other sensitive areas such as green belts and areas of outstanding beauty.

The planning of rural settlements has broadly adhered to the 1942 recommendations of the Scott Report. This committee was appointed in 1941 to consider the conditions which should govern building and other constructional development in country areas. It viewed rural housing as a 'serious problem', contributing significantly to the drift from the land.

> In spite of the falling numbers of the rural population, there was a growing shortage of housing accommodation in many districts, largely due to cottages being rented by others than farm workers. Over the country generally, many rural workers were living in cottages which should have been condemned as uninhabitable.
>
> (Scott Report 1942, 17)

The committee regarded 'the improvement of rural housing as an essential prerequisite to the re-establishment of a contented countryside' (p. 48) and laid stress on the building of cottages 'as near as possible to the village or nucleated settlement' to relieve the housing shortage (p. 50). This recommendation that new building should be concentrated in and around the existing settlements has been a major influence on post-war policy for housing and planning in the countryside, and it is worthwhile to quote the reasons advanced by the Scott Committee (1942, 73):

> Whether or not there is to be any considerable influx of industry and industrial population into country areas there must inevitably be a good deal of new building to replace old outworn cottages, farm houses and farm buildings which are now such common features in most districts as well as much completely new housing. It is inevitable, too, that there will be new week-end cottages for townsmen, new hostels, holiday camps, camp schools and so on; new petrol-filling stations, garages, restaurants and hotels for the traveller, and especially new houses and bungalows for the pensioner and the retired.... The farm worker and his family have far more chance of a happy social life and better opportunities of

> developing as self-reliant and responsible members of society if they live in a village. This is true of all dwellers in the countryside. It applies to the week-ending townsman and to those people who now go to live in ribbon developments as well as to genuine countrymen. Though not all country dwellers can live in groups, planning schemes should be so designed as to direct all new settlers into country towns and villages except where they can advance some decisive reason why they should be housed in the open countryside. Such a direction would not only benefit the settlers themselves, it would go a long way towards putting an end to that sporadic and scattered building which has done so much to spoil and suburbanise, and consequently (as many witnesses have pointed out) to affect seriously the agricultural production of large parts of the country during the last two or three decades, particularly those parts within the immediate sphere of influence of the towns. Closer building will also facilitate the provision of services.

The arguments put forward in justification of the post-war containment of rural settlements thus include the protection of agricultural land and the beauty of the countryside, the lower costs of service provision, and the alleged social benefits of domestic proximity. The Scott Committee never recognised the apparent conflict of objectives arising from its proposals, calling on the one hand for new building to relieve the serious rural housing shortage, and on the other for agricultural land to be protected from all but the most essential developments, with the onus placed on developers to show that it was in the national interest 'for good land to be alienated from its present use'. It may be that the Scott Committee intended no constraint on the numbers of new houses built in the countryside, but only to constrain their location: in practice these are so closely interrelated through the land market as to be inseparable.

Given the much greater strength of the farming and conservation lobbies, it is perhaps not surprising that the recommendations for protecting agricultural land and amenity were seized upon while the arguments for new cottages were quietly forgotten. In Newby's (1980, 239) view, this outcome

> was the product of an unholy alliance between the farmers and landowners who politically controlled rural England and the radical middle-class reformers who formulated the post-war legislation. The former group had a vested interest in preserving the status quo, while the latter, epitomised by the nature-loving Hampstead Fabian who enjoyed country rambling at the week-ends, possessed a hopelessly sentimental vision of rural life. The rural poor had little to gain from the preservation of their poverty,

> but they were without a voice on the crucial committees which evolved the planning system from the late 1930s onwards. Consequently the 1947 Act framed the objectives of rural planning in terms of the protection of an inherently changeless countryside and a consensual 'way of life' which overlooked important social differences within the rural population.'

The question of which groups had a vested interest in preventing new housebuilding in the countryside is closely related to Weberian notions of housing class, already touched on in Chapter 2. This highly significant issue is returned to in Chapter 4 in discussing rural housing classes.

Therefore, the Town and Country Planning Act 1947 laid the basis for a post-war policy of tight controls on rural housebuilding, which have steadily reduced the amount of countryside taken for housing from 25,000 hectares a year in the 1930s to 5,000 today. Its operation has been summarised by Rogers (1976, 96) as follows:

> The planning machine which was set up from 1947 put a tight control on new housing in rural areas which has meant that, far from encouraging rural authorities and private enterprise to build houses as in the 1930s, there has been active discouragement, particularly with regard to building in the open countryside, except where it can be proved that new housing is necessary for essential agricultural workers. The twin objectives of urban containment and countryside preservation which have been explicit in land use planning since the Second World War have inevitably had repercussions for rural housing.

This policy has been communicated through successive government circulars in 1950, 1960 and 1969 which have directed local authority planners to refuse permission for new houses in the countryside, unless a special need existed. Even within villages, development unrelated to traditional rural-based activities was to be discouraged: carefully sited infill development and perhaps some limited 'rounding-off' at the edges might be permitted, but this was often carefully controlled by the imposition of an inflexible village envelope (Blunden and Curry 1985, 88). Even where village envelopes are no longer fashionable, tight restrictions on new housing development outside key settlements are the general rule in current structure plans (Derounian 1979).

The economic and social consequences of restricting the supply of land for housing are likely to have been profound. In the first place, there is likely to have been an effect on the price of housing, and on the price of land itself. Because of its importance, this relationship is discussed both in relation to theory and then, in relation to the empirical evidence available.

There are two schools of thought as to the relationship between land prices and house prices. The first draws on Ricardian rent theory to view land prices as a residual (e.g. PIEDA 1986), whereas the second views both land price and development gain as elements of surplus value to be contested between the landowner and the developer according to their respective market strengths (e.g. Ball 1983). In essence, the Ricardian approach views landowners more or less as price takers, whereas the alternative approach emphasises the struggle over the conversion of the gross development profit into land prices between developers and landowners. Each of these approaches is now elaborated.

Ricardian rent theory views land as a factor of production, whose value derives from the value of its final product. Thus, the value of farmland derives from the value of agricultural output, and the theory would predict that a fall in expected agricultural output prices would lead to a fall in the value of farmland. In relation to housing, an increase in the level of house prices will lead developers to seek to expand their output, and therefore to purchase more land on which to build: this increase in the derived demand for building land will then force up the price of land for housebuilding to the point at which it is only just profitable for the builder to obtain the land. House prices therefore drive land prices, rather than land prices impacting on house prices.

The extent to which any increase in house prices is captured by land value rather than by the other factors of production depends upon the price elasticities of the factor inputs, with the greatest share of the increase being captured by the least price elastic factor. The supply of land is price inelastic according to Neuberger and Nicol (1975), and Drewett (1973) has argued that 'the supply of other factors that must be combined with raw land to produce houses is relatively elastic'. It follows that most of the variability in house prices falls upon the price of housing land.

The Ricardian view that house prices drive land prices, rather than vice versa, does not, however, preclude the availability of housing land affecting house prices. But the causality is more complex than often suggested by housebuilders. If the supply of land for housebuilding is restricted, either by planning policies or by ownership constraints for example, there is no direct effect on the price of housing land. Under the assumptions of Ricardian theory housebuilders are already paying the highest price compatible with profitable use of the land. However, builders will be unable to obtain as much land as they wish for housing, and construction will consequently fall in time, leading to an eventual reduction in the supply of housing. House prices will therefore rise, and as they rise builders will be able to increase the price offered for housing land. Thus both house prices and land prices will have risen, but again it is the increase in house prices which drives the increase in land prices even though the original cause was a restriction on land availability. It is perhaps also worth noting that if the restriction on land supply was due to landowners' unwillingness to sell, one might expect that the increase in land prices would encourage the release of further land for house-building: if the restriction derived from planning policy, then it is unlikely that the rise in land prices would bring forth any increase in the amount of land available for building.

Consideration of the role of landowners may call into question the view of landowners as price takers, implicit in the orthodox Ricardian analysis. Evidence was found by Drewett (1973) that landowners did not adopt such a passive role:

> The land buyer uses land as a stock-in-trade and at some stage he must go out and buy land or go out of business. The landowner, on the other hand, views land differently; he may not want to use his land and with expectations high on a rising wave of value he need not sell.... As there is no tax on holding land, the disincentives of continued ownership are minimal.

This analysis, since developed by Lichfield and Darin-Drabkin (1980) and Evans (1983), emphasises the monopoly power of landowners, given the uniqueness of each plot of land. It is suggested that landowners can withhold their land from the market until, through the process outlined, a housing shortage is created and the price rises to a level acceptable to them. Whether landowners in fact have this power, and employ it, is an empirical question rather than a theoretical one, and Munton and Goodchild (1985) found no evidence to support this suggestion.

Ball (1983) agrees that the orthodox Ricardian analysis, rather than its variant, is most consistent with the empirical evidence; but he argues that while 'its hypothesis may fit the data, it fails to answer the vital question of how housebuilders have managed to avoid landed property exercising a monopoly power over land price'. He therefore proposes an alternative theory of land price determination, in which the land price is the outcome of a struggle between developers and landowners over a share of the gross development profit or development gain. The implication of Ball's analysis is also that land prices do not affect house prices; moreover, house prices do not have the deterministic influence on land prices indicated by Ricardian theory, since the price paid for land reflects the struggle between landowners and developers over development gains. Planning controls themselves are important not so much in terms of any effect on the overall level of house prices, although there may be such an effect, but rather in their distributional impact on the relative market positions of landowners and developers and hence, on the share which land prices take of development gains.

While both Ricardian theory and Ball's analysis would admit that planning restrictions on land supply might restrict the supply of new housing and so lead to increases in house prices, it is important that the correct direction of causality is understood: 'for example, the view that land prices push up house prices could lead to an argument for land price controls – the same does not follow from the view that house prices push up land prices' (PIEDA 1986), Whether planning restrictions on land supply have in fact restricted the supply of new housing and thus forced up house prices is an empirical question, however.

Although the data on housing land prices are rather meagre, Ball (1983) has examined data on development profitability and on land prices from 1970–81 to demonstrate the direction of the causality between house prices and land prices, leading him to the conclusion that

> there is a remarkably close lagged relationship between changes in development profitability and residential land prices. It actually appears that when their profits rise builders do buy more land, forcing up its price, and when profits fall they buy less, apparently validating the Ricardian residual view.
>
> (Ball 1983, 113)

Ball also notes that

> if the residual view of land price is correct the proportion of house price which corresponds to current land price should also vary in line with the building cycle. During periods of rising development profitability the proportion of house price represented by that profit will rise, and vice versa. So the share of house price attributable to land prices should also rise and fall with development profitability booms and slumps.
>
> (Ball 1983, 113–14)

Again, his data shows that this has generally been the case during the 1960s, 1970s and 1980s, so supporting the Ricardian thesis.

However, Ball contends, on the basis of data for the period from 1970 to 1982 only, 'that land prices have not risen in the long run as a proportion of house prices as would be expected if land shortages were the principal cause of the long-term upward trend in house prices'. This is contrary to the conclusions of Drewett (1973) based on his analysis of 1960s data: 'land prices have increased steadily since 1963 and more rapidly after 1966, when the median value nearly doubled in many areas. This is reflected in the increasing proportion of land as a cost in the final house price.'

Data from the DoE for the period 1963 to the present confirm that there has been no simple and consistent trend in average land price as a proportion of new house prices. This proportion rose from 1963 to 1966, from 1967 to 1973, and from 1977 to the present, with falls in 1967 and during 1973 to 1977. But, as noted, one would not expect to find a stable trend from year to year given the association between this proportion and the booms and slumps in gross development profitability. Rather, one might expect a long-term, underlying tendency for land prices to rise as a proportion of house prices if land shortages were a principal cause of long-run house price increases. Price movements over a single decade are clearly inadequate for the detection of such a tendency. Even over the twenty-three year period for which data are available from the DoE any long-term trend is small in relation to short-term variations about it: the application of simple linear regression indicating that there is an upward trend, amounting to an increase of one percentage point in this proportion every six years, is not therefore conclusive. Nevertheless, it is probable that land prices are increasing in proportion to house prices when one considers that plot sizes have decreased substantially over this period, indicating that there has been a greater rise in the price of land per hectare in relation to house prices, than in the price per plot.

One study which has attempted to take a truly long-term perspective is that of Hallett (1977). This relied on Vallis's series of index values between 1892 and 1969, based on auction sales data prior to 1964. These data do therefore need to be treated with caution, particularly since they derive from only a small number of observations. The price of residential land was found over the whole period to have increased by substantially more than retail prices, average earnings or other property sectors; and under the post-war planning system the value of residential land had risen more than threefold in real terms. Despite this, Hallett concluded that land values had grown no faster than inflation. 'Yet the land price data he employs, over the time period he uses, do not seem to support his position. Land values have risen in real terms during the 20th century' (Munton and Goodchild 1985).

Furthermore, the most probable cause of such a trend is a supply constraint, such as planning controls. While short-term fluctuations in house prices are most likely to be caused by variations in effective demand, longer-run structural movements in house prices are more likely to be caused by supply-side factors, since otherwise producers would increase their output in areas of high prices to meet demand, so ultimately reducing house prices again. 'To put it simply, demand shifts can cause prices to diverge for quite lengthy periods within and between housing markets but relatively 'high' prices will persist only if supply is constrained in some sense' (PIEDA 1986). Such constraints on supply might include labour shortages, shortages of materials or restrictions on the availability of credit: but the most likely long-run constraint on the supply of houses is a shortage of housing land, either because of the inherent limitations of space (e.g. in a city centre) or because of land-ownership or planning controls.

While the inadequacy of data precludes any attempt to undertake more sophisticated tests of the hypothesis that land prices have risen as a proportion of house prices over the long run, there are other relevant hypotheses deriving from the above theoretical discussion which may be tested. In particular, if land supply constraints were not a factor in causing high house prices we should expect land prices to occupy a greater proportion of house prices in more central urban areas than on the rural fringes of such areas, since the supply of houses ought to be more constrained in the more central locations by the inherent limitations of space. In fact, that proportion is highest in the outer metropolitan area (Shucksmith and Watkins 1988c): this suggests that land supply constraints are effective (together with buoyant demand) in creating a scarcity of housing and high house prices in the outer metropolitan area at least.

The suggestion that planning restrictions on land supply are a factor of importance in raising house prices is supported by further evidence at a much more detailed level collected in Scotland (PIEDA 1986). PIEDA argued that

> if land supply constraints are affecting the operation of the housing land market and the housing market, we should expect to observe: (i) substantial variations between areas in the level of land prices; (ii) the association of high land prices and high house prices.

The analysis found evidence to confirm each of these hypotheses. PIEDA found considerable variations between areas in the level of land prices, both between Edinburgh and other cities and also around Edinburgh. A very close association between land prices and house prices was also found, although most of the variation in house prices was accounted for by the attributes of the house. Most interestingly, the study found that the areas of higher land prices tended to have lower levels of land with planning permission per capita: yet if the development process was adjusting smoothly in the absence of any supply-side constraint, 'the amount of land per capita with planning permission would vary from area to area depending on the rate of development but differences in the figure would not indicate areas of relative land shortage – supply would be coming forward to meet the appropriate level of demand.' The evidence therefore implied that supply was failing to respond to demand, partly because of planning restrictions. 'In brief, the statistical evidence is consistent with the hypothesis that, in some areas, restrictions on land supply, related to the planning system, have led to an increased level of house prices and land prices' (PIEDA 1986).

In terms of housing provision for lower income groups in rural areas, this increase in housing land prices has had several important results. First, land within villages has become more expensive for local authorities and housing associations to buy, and so fewer council houses have been built in rural areas. Rogers (1976, 114) quotes local authority estimates in Oxfordshire of the annualised cost of building and servicing a new council house in 1974–5 of from £19 to £27 in the villages compared to £6 in Oxford itself. Second, higher land prices have tended to encourage more expensive private developments both on infill sites and on those new estates adjacent to existing settlements which have been approved: 'put briefly, much private development is of a type, and consequently at a price, which effectively excludes many of those rural people who have the greatest need for housing' (Rogers 1976, 116).

The social dimension of these outcomes is elaborated by Newby (1980, 187):

> As prices inexorably rise, so the population which actually achieves its goal of a house in the country becomes more socially selective. Planning controls on rural housing have therefore become – in effect, if not in intent – instruments of social exclusivity.

Hall *et al.* (1973, 406–9) examine the distributional effects of the containment of settlements on several specific groups. They conclude that the major gainers are the wealthy, middle-class, ex-urbanite country dwellers and the owners of land designated for development. The losers include the owners of land on which development was not permitted, the less affluent among the new suburbanites, the council tenants and tenants of private rented property. Summarising, they argue that the effects have been regressive in that 'it is the most fortunate who have gained the most benefits from the operation of the system, whilst the least fortunate have gained very little' (p. 409).

In general, therefore, land use planning has pursued the objectives of landscape protection and the protection of agricultural land at the expense of the objective of rural housing provision. Not only has this had regressive distributional consequences, as described by Hall and Newby among others, but as argued in the discussion at the beginning of this chapter it is likely that the outcome has been inefficient as well, in that it has contributed to the decline of rural communities and so to undesired consequences for landscape protection and public enjoyment of the countryside. The distributional aspects of this policy outcome and its class basis will be discussed further in Chapters 3 and 4.

There are, however, four specific policy measures which have been adopted in an attempt to ameliorate this conflict of objectives, and these will now be discussed in turn. The first two relate to the construction of and the tenure of farm cottages. The third is the 'rural safeguards' amendment to the council tenants' right to buy. The fourth, which has particular relevance to this book and which will therefore be examined in more detail than the first three, is the 'local needs' policy.

Farm Cottages

Farm workers' cottages have been exempted from two of the most pervasive policies affecting rural housing opportunities since the war; presumably because of the importance afforded to the strengthening of post-war agriculture and because of the political force of the farmers' lobby. In the first place, the construction of a new farm cottage in the open countryside may be permitted despite the general restrictions on new building if an essential agricultural need can be established. Development Control Policy Note 4 (1969) explains that

> a person who wants to build a farm dwelling in a rural area must produce evidence of need to offset the general planning objections to such development. Unless real need can be established the normal planning considerations will prevail.

The DoE's Circular 24/73 elaborates the meaning of need in this context, making it clear that it means agricultural need and not housing need.

> Unless the applicant shows that there are valid reasons why a dwelling house should be erected on the farm rather than in a nearby village, the normal planning considerations will apply and the need for agricultural dwellings should be met as far as possible by building in an accessible village, hamlet or existing group of dwellings.... The following factors will be material in assessing whether there is agricultural need for a farmhouse or cottage on the farm: (a) the viability of the farming enterprise; (b) the labour requirements of the enterprise (c) how many workers will need to live on the farm; (d) the existing accommodation on the farm and the reasons why it does not meet the needs shown in (c) above. Need in this context means the need of the farming enterprise rather than that of the owner or occupier of the farm.

The planning authority generally asks the Ministry of Agriculture (MAFF) to advise them on the presence of agricultural need, and if satisfied that a genuine need exists the council then normally grants planning permission subject to a condition that the house will be occupied by a person employed locally in agriculture. Such a condition will only be removed on a subsequent application if it can be shown that 'the long-term needs for dwellings for agricultural workers, both on the particular farm and in the locality, no longer warrant its reservation for that purpose' (Circular 24/73, section 14). Dunn *et al.* (1981, 203–4) confirm that this policy is applied by most planning authorities, and observe that 'in this way agricultural dwellings receive preferential treatment within the planning system'.

The significance of this preferential treatment is that it appears, on first sight, to be an exception: the only instance whereby the protection of landscape and agricultural land has taken second place to the housing needs of the rural workforce. In reality, however, as circular 24/73 makes clear, this preferential treatment has nothing whatever to do with housing objectives but is simply another manifestation of the privileged position afforded to the agricultural industry since the war. Isolated farm dwellings are permitted not where farmworkers' housing needs are particularly acute, but where they are essential to the farming enterprise and its contribution to expanding post-war agricultural output.

Nevertheless, there is an analogy with our earlier discussion. Both the protection of agricultural land and the provision of essential workers' cottages are necessary to the objective of expanding agricultural output, and yet they conflict. Equally, it has been argued that both the protection of landscape and the maintenance of rural communities are vital to the long-term preservation of the countryside's amenity, despite the apparent conflict over land for housing. Perhaps the contribution of rural housing provision to landscape preservation is merely less obvious than the contribution of farm cottage provision to agricultural production; an equally convincing explanation, however, is that the interests of landowners, farmers and home owners have prevailed in each case.

The other respect in which farm cottages have received special treatment in post-war policy has been in their exclusion from the Rent Acts. Until 1976 in England and Wales, and still today in Scotland, farm workers living in tied cottages had little or no security of tenure. As Gasson (1975, 92–6) has described, the legal position varied slightly according to whether the farm worker was a service tenant, a service licensee or an ordinary licensee, but in all cases the farmer was virtually guaranteed repossession if the worker lost his job. As G. Clark (1982, 31) has said,

> Whichever category he was in, he could find himself jobless and homeless in quick succession unlike most other tenants. Since rural areas usually have fewer council houses than the cities and house purchase was normally beyond the means of low-paid farm workers, the potential for distress was considerable.

The Rent (Agriculture) Act 1976 extended security of tenure to virtually all workers in agriculture and forestry in England and Wales (Rossi 1977). On ceasing his employment the worker now becomes a statutory tenant with full security of tenure under the Rent Acts: if the farmer requires the dwelling for another employee, however, and the agricultural need is proven, then the local authority is required to use its best endeavours to rehouse the tenant. Both the tenant and the farmer seem to have gained by this legislation (Dunn *et al.* 1981, 218) at the expense of the local authority, which now has to face the additional burden of rehousing, and of those on the council waiting list who find themselves overtaken in the queue for council housing. While this issue is highly germane to the theme of differential access to rural housing (and is discussed in relation to that theme in Chapter 5), it is not of major relevance to the conflict between countryside objectives and housing objectives, and is not pursued further here.

Rural safeguards on council house sales

A more recent government recognition afforded to the existence of specifically rural housing problems came in the form of rural safeguards attached to council tenants' right to buy. The Housing Act 1980 and the Tenants' Rights (Scotland) Act 1980 conferred on council tenants the right to buy their homes at a discount, which has averaged 44 per cent (Foulis 1987). Rural council houses are particularly attractive to purchasers, who may resell at a considerable profit. Phillips and Williams (1984, 329) have shown that sales in the smallest districts of England were at roughly five times the level of sales in large cities. More detailed research in the South Hams District of Devon, relating to voluntary sales prior to the 1980 Acts, has shown that council house sales tend to be greatest in the more rural parishes (Phillips and Williams 1982). These findings have been confirmed in a number of other studies, in areas as diverse as Gordon District (Williams and Sewel 1987), West Norfolk (Friend 1980) and East Hampshire (D. Clark 1982). The studies have also shown that sales are generally heaviest among the best stock, which is sold mostly to middle-aged tenants, leaving a council stock which is depleted not only in quantity but also in quality.

In Scotland, Table 3.1 shows clearly that council house sales in rural areas have been much heavier than the national average: between 1979 and 1986, roughly half the rural districts in Scotland had lost over 14 per cent of their stock, compared with a Scottish average of 7.0 per cent of stock sold. Within these local authority areas the landward stock of council houses is most at risk: if all early applications had proceeded in Banff and Buchan, for example, the stock of council houses would have fallen by 6 per cent in Peterhead, in more rural Turiff by 13 per cent, and in the small village of Gardenstown by 25 per cent, up to 1988 (Housing Plan 1983–8).

Table 3.1 Council houses sold in Scotland, October 1979 to September 1986 (In percentages)

Northeast Fife	21.0	Edinburgh	9.2
Gordon	18.1	Aberdeen	7.5
Stewartry	18.1	Dundee	3.9
Badenoch and Strathspey	18.0	Glasgow	2.6
Western Isles	17.7		
Nairn	16.1	Scotland	7.0
Kincardine and Deeside	15.6		

Source: Hansard, 24 February 1987, cols.188–9

The effect of council house sales, in rural areas as elsewhere, is now fairly well established. While a local authority suffering from reductions to its capital allocation may be able to use capital receipts from sales to bolster its building programme (more so in Scotland than in England, where only 20 per cent of such receipts may be spent), these receipts are far too low to allow the replacement of the stock sold. Further, in the long term the local authority will lose more in rent foregone than it receives from the sale (Kilroy 1982). As a result of these financial effects, there will be a gradual reduction in the number of relets (estimated by the House of Commons Environment Committee to be a loss of 2.6 per cent p.a.), to the detriment of those on council waiting lists in future years. English and Martin (1983, 102) attempt to summarise the consequences:

> Council house purchase at a discount is obviously advantageous to the buyer, but it seems likely to be costly to the community as a whole. As well as diminishing the quality of the public sector (and in some rural areas going a substantial way towards eliminating it altogether), the cost in housing subsidies to the taxpayer will be increased. No serious commentator would now claim that sales will save public money.

Local housing authorities in rural areas have resisted mandatory council house sales, arguing that they will lead to 'a greater pressure on those whose only real hope of a house is through the local authority' (WIIA 1980) and will mean 'frustration of the hopes and desires of the number of applicants on the waiting list' (Gordon District 1978).

During the passage of the acts considerable pressure was exerted by rural councils and other pressure groups who demanded safeguards for rural areas in recognition of the high level of sales expected and the danger of resale to retirement migrants and second-home purchasers. The unlikely alliance of Rural Voice, Shelter, the NFU and Plaid Cymru achieved some success in this respect in England and Wales. In national parks, AONBs and specially designated rural areas, while the right to buy remains, the council must either be allowed first refusal on resale during the first ten years or the council can attach a covenant limiting resale to people who live and work in the locality (Housing Act 1980, section 19). Very few rural council houses are within these areas, however: out of more than 130 districts in England which applied to become designated rural areas only eighteen were actually approved, apparently on the basis of a high incidence of second homes. These safeguards were further criticised by the House of Commons Environment Committee on the grounds that the locality condition relates to the purchaser rather than to the occupier (so the house can be used as a holiday cottage); that resale may be restricted to local residents, but that these will probably be the more affluent; that the ten-year period of first refusal is too short; and that most authorities will be unable to afford to buy back the properties at full market value having sold at a discount (Phillips and Williams 1981).

In Scotland, the political pressure for rural safeguards was less. Thus, 'Scottish Ministers rejected the case for restrictions, arguing that the demand for second homes was less in rural Scotland', although a token concession was eventually made 'to satisfy their own supporters' (Gillett 1983, 20–1). This was that once more than a third of a district's council house stock has been sold, and the secretary of state is satisfied that an unreasonable number of these have been resold as second homes, he may then give the council the power to impose a ten-year right of first refusal on resales of stock sold subsequently (Tenants' Rights (Scotland) Act 1980, section 4). No area in Scotland has yet been designated in this way, and since no mechanism exists to monitor resales it seems unlikely that the Scottish safeguards will ever be implemented. As the minister has stated, 'there is no evidence to suggest that ex-council houses which are re-sold are doing anything other than making a valuable contribution to the range of owner-occupied housing available to local residents' (SDD 1986), but neither is there any evidence to the contrary.

Nevertheless, the most significant aspect of the acts for this chapter is their recognition of specifically rural problems of access to housing in areas of high landscape value (national parks and AONBs) and areas of high second-home ownership (designated rural areas). It may be the case that the rural safeguards in the legislation are ineffective, but they are a recognition of the wider rural housing problem and of the importance of the council stock in meeting rural housing needs.

Local needs

The English restrictions on council house resale also embrace the argument that those who live and work in the locality merit preferential treatment. As Rogers demonstrates, this concern for local housing needs is also widespread among rural local authorities, as reflected in many structure and local plans as well as in the residential qualification often operated for access to council house waiting lists.

> The growing concern for the housing needs of local rural people has thus been reflected by both local and national government and enshrined not just in practice but in legislation. At the local level the desire to protect local housing markets by structure plan policies and by planning practice has resulted in what Clark has with some justification called 'one of the most radical attempts in Great Britain to intervene in a rural housing market'. [The subject of the case study in Chapter 6.] At the national level, the recognition of the special needs of some local rural people has caused a significant departure from a strongly held desire on the part of the Conservative Government to allow unrestricted council house sales.
>
> (Rogers 1985a, 374)

More than twenty structure plans have explicitly declared their intention of operating local need policies (Rogers 1985a, 371) and yet it is not at all clear what is meant by this term: indeed, the secretary of state is said to have dismissed local need as a misleading and impracticable concept (Derounian 1980, 87). Elson *et al.* (1979, 3) comment that:

> One of the most striking findings of our work to date has been the lack of definition and vagueness which surrounds local needs policies. There are three questions which must be addressed when considering the issue of the definition of local needs:
> a) What is the definition of need?
> b) What is the definition of local?
> c) Can any definition of the above be combined in any meaningful sense?

Such comments were also made by Rawson and Rogers (1976, 23) in an earlier review of the rural housing content of structure plans:

> The issues of local need are emotive and popular at the present time, yet there is little attempt to define what is really meant. The nearest approach to a definition is probably made in Hertfordshire, which includes households resulting from natural increase in the existing population, plus the relief of homelessness, reduction of overcrowding and the replacement of substandard housing.

Frequently, policy statements explicitly or implicitly exclude any provision for migrants who will work for employers new to the area or who will commute out of the area to work (Buckinghamshire), or 'for second, retirement and holiday homes and commuting [since] there is no justification for new housing to be built specifically for these purposes' (Cumbria County Council 1980, 53). Norfolk Structure Plan (1977, section 3.3.14) is quite explicit:

> Although the primary function of the Structure Plan in relation to housing is to ensure that sufficient land is allocated in appropriate places to meet housing needs in the county, it is not accepted that these allocations must also be sufficient in all areas to meet the local housing demand.

Housing demand, particularly from in-migrants, has been a significant factor in the failure of local authorities to meet housing needs; but instead of increasing targets to meet both normative need and demand, the unlikely assumption is made in structure plans that external demand will be somehow suspended so that needy locals will no longer be outbid for the existing private stock and for the limited new development permitted. Instead, local needs are less likely to be met as a result of such policies, since competition between locals and incomers will be intensified by the reduction in new building: the price of houses will be bid up further beyond the means of lower income groups, reinforcing the tendency for those with unsatisfied housing needs to be outbid.

Suffolk County Council's (1977) Structure Plan's recognition of this difficulty is a rare exception to the general failure of planning documents to foresee the market consequences of planning policies:

> The listing of large numbers of minor centres and small villages where housing development is to be related to the needs of the local community cannot of itself ensure that the housing permitted will become available to local people. The operation of a free market in land and housing means that some at least of the houses in these villages will be occupied by people from outside.

Unless some mechanism ensures that houses intended for local needs actually go to those accepted as in local need, it is inevitable that restricting development to local needs only will fail to help. This is because local needs policies implicitly rely upon the suspension of external demand, and yet in practice such policies merely restrict supply. As Murie (1976, 56) has argued, 'a plan to meet certain needs must incorporate the method of channelling resources to those in need', and it is clear that local needs policies generally lack any such mechanism. However, there does appear to have been a gradual acceptance by both ministers and local planners during 1988 and 1989 of the desirability of releasing land on which housing associations may build social housing for locals, and this differs from previous local needs policies in that it does incorporate a mechanism for channelling resources to those in need. This crucial aspect is discussed further in Chapter 7.

In summary, then, it seems that the term 'local needs' has been used in a loose way, to convey only a very imprecise and superficial concern for the welfare of local people, within the context of general restrictions on housing development. Policies have been formulated to meet these vague, undefined local needs in ways which are unclear and unstated. The concept of need, and the validity of local claims to rural housing, are discussed further in Chapter 4. Elson's view is that 'local needs' is a deliberately vague and imprecise term, offering a politically convenient umbrella description of a conglomeration of local objectives, rather than resting on any formal conception of 'need' itself:

> The term 'local needs' has come to represent, in such areas, a decision upon a politically acceptable level of growth and the phasing of development. Complete or rigid restraint is seen as unworkable and unacceptable by central and local government and has been resisted by many district councils as well as wider housebuilding and industrial development interests. Local needs policies are an encapsulation of a wide variety of politician's and planner's concerns.
>
> (Elson 1981, 63)

Local needs policies can therefore be interpreted as a further attempt to resolve the conflict of objectives between restraining development in the interests of landscape and farmland protection and encouraging housing provision for the benefit of rural communities. Such policies can be seen either as a well-intentioned, but misguided, attempt to reconcile these inherently conflicting objectives of public policy; or they may be viewed as attempts to legitimate tighter restrictions on development by apparently meeting the interests of several housing classes, while obscuring the fact that they serve the interests of owner-occupiers and others, and frustrate the interests of non-owners in particular. This theme is developed in greater depth in the case study in Chapter 6.

Conclusion

This chapter has examined the conflict of objectives inherent in policy towards rural housing in Britain. A neo-classical economic perspective on the objectives of countryside policy and of housing policy suggested that, on the one hand public intervention is justified to encourage the provision of the public good of landscape protection. On the other hand, intervention is required to promote the welfare of rural communities both on the grounds that this is a form of producer good essential to the production of two public goods, landscape protection and public enjoyment of the countryside, and to correct for housing market imperfections.

However, in practice, rural housing policy since the war has been dominated by land use planning considerations directed towards containing settlements and protecting landscape and farmland, at the expense of rural housing provision. This policy was examined from a political economy perspective, and was seen to have resulted in regressive distributional consequences as well as being an inefficient allocator of resources, in that the decline of rural communities may ultimately threaten the landscape's preservation. Further, market situations in relation to land release for housebuilding were seen to be the basis of class conflict between owner-occupiers and non-owners of rural housing. Four policy measures which acknowledge this conflict of objectives and which reflect the underlying class conflict were discussed, but while policies to ensure that farm workers could be housed near their work were effective, policies designed to safeguard the stock of former council houses for local needs and planning policies intended to discriminate in favour of local housing needs were largely unsuccessful.

The next two chapters turn away from the theme of economic efficiency and conflicting objectives, and explore instead the other main theme of this book, the equity aspects of differential access to rural housing. We have seen that the distributional consequences of current policies are regressive in general terms, but this requires a more detailed analysis. Further, the conflict between housing provision and landscape protection has been proposed as the basis for a model of rural social stratification, allowing Weberian housing classes to be identified according to their market situation in relation to land release for rural housebuilding, and such a model requires elaboration. Chapter 4 therefore explores briefly the concept of need and the basis of local claims for priority in rural housing allocation; it then develops a concept of rural domestic property classes and proposes two typologies of rural housing classes, one relating to the accumulative potential of rural housing and one relating to housing consumption. These are then used in Chapter 5 as a framework for considering distributional outcomes, and for understanding the forces which give rise to unequal outcomes.

Notes

1. This is not to argue that the objectives attributed to government by economists are necessarily those which politicians rationally act upon: on the contrary, the pre-eminence of urban containment reflects politicians' susceptibility to interest groups.

Chapter four

Rural housing opportunities, housing classes and status groups

The concept of need

Inequity and need

It was stated at the outset that one of the two themes of this book is the analysis of differential rural housing opportunities and the identification of the incidence of housing disadvantage. According to Robson (1979, 71), inequity is inevitable: 'no level of housing provision is likely to dampen the feelings of deprivation and unfairness so long as housing is produced and distributed within a market economy in which money rather than need determines access to housing.' But what is meant by need? And how does this relate to differential housing opportunities and to equity?

Implicitly, inequity means a failure to allocate resources according to the guiding principle of need. Unsatisfied needs may result, therefore, either from inequitable access to housing resources, or from insufficient total resources being devoted to housing to satisfy all needs. For even if all housing resources were distributed according to need, they are unlikely to be sufficient to meet all needs, if only because acceptable standards will rise as more resources become available.

The existence of unsatisfied needs for housing in rural areas is the primary concern of those who see rural housing as a merit good. Successive governments have espoused the 'traditional aim of a decent home for all families at a price within their means' (HMSO 1971, 1; HMSO 1977, 7) and this would seem to apply to rural families as to others. Those without 'a decent home', however defined, are those whose needs are unsatisfied. The idea of 'a decent home' is necessarily a changing one, as society's or the individual's expectations increase: any analysis of which households remain without 'a decent home', and why, is a study of inequitable access to housing and of resource allocation. As Williams (1974, 71) has argued, any researcher who accepts this notion of housing as a merit good is 'forced to unravel the same tangled skein of conflicting roles and judgements that the "needologists" have been grappling with'.

A discussion of the concept of need is therefore a necessary preface for any analysis of public intervention in housing provision, and it may help to clarify the relationship of need to inequity of access. The distinction between need and effective demand is fundamental. Demand is an essentially descriptive term, used to denote the relationship between the price of housing and the quantity and quality of housing for which people are able and willing to pay. In contrast, housing need is a prescriptive term, used to denote the inadequacy of existing housing provision when compared with some desired norm, presumably 'a decent home'. The prescription is that provision should be improved to at least attain that norm. Thus, a household may be thought by itself or by others to 'need' a home, but be unable or unwilling to pay the market price: in such a case no demand exists.

This raises the question of who is to determine whether or not need exists. Need, like the concept of a merit good, is necessarily a subjective phenomenon, and opinions are likely to differ in any particular case. As Culyer (1976, 44) has noted, 'the making of value-judgements...is the very essence of establishing the meaning of the word "need"'. In the context of rural housing, Shucksmith (1981) adapted Bradshaw's (1972) taxonomy of need to explain why the methods most commonly used in rural areas to estimate housing need tend to be unreliable.

Need as measured by the council waiting list, for instance, concerns only the overlap between normative need and felt need and then it is only admitted if the need is actually expressed. That is, both the individual applicant and the local authority managers must agree that need exists, and an application to join the list must be completed before an individual's name will appear on the council's list. This measure may exclude many households who feel themselves to be in housing need, because they are not admitted as such by the local authority managers; and it excludes many others who the council would recognise as in need, because their need is not expressed. Nevertheless, this method still appears to be the major source of information on which rural authorities' housebuilding is based, as recommended by the SDD (1978). The extent to which need is felt but not expressed, whether according with normative criteria or not, can only be estimated by more subtle techniques, such as special surveys.

The demographic approach of the Housing Investment Programme (HIP) returns in England and Wales, analagous to the Housing Plan in Scotland, measures only normative need as defined by central government. The total number of existing households is added to estimates of the numbers of concealed and involuntary sharing households, households accepted as homeless, and overcrowded households (collectively termed 'potential households'). This total is then compared with the total number of dwellings, minus vacant dwellings and second homes, to give the dwelling–household balance. Since data concerning most of these categories are out of date (1981, or even 1971 census) or non-existent, and because definitions of such categories as involuntary sharing households vary so widely, the resulting figure can only be a very rough guide to the extent of normative need at district level. And as Murie (1976, 54) has noted, 'these limitations become more pronounced when the estimates become, as they must do for planning purposes, projections or predictions'.

Hidden needs

No reliable estimates therefore exist of the extent of rural housing needs, either in terms of felt needs or of normative needs. Measures of expressed needs, in the form of council house waiting lists, may be analysed but will certainly underestimate felt or normative needs for several reasons.

Several categories of need are not admitted by councils. Larkin (1979, 72) has suggested that 'such restrictions tend to be more common and draconian in rural than in urban areas'. Thus, tenants of tied houses in Scotland, those in winter lets, caravan dwellers and owner-occupiers are commonly excluded (SDD 1979; Shelter 1982). An SCSS assessment of housing plans in Dumfries and Galloway (Smith 1979) noted that Stewartry District, for example, guards against 'over-provision' for what it sees as 'an insatiable demand for public-sector housing' by restricting the criteria by which an applicant can enter the priority need section of the waiting list. Others are regarded as 'not truly in need'. Similar practices were found in the Lake District (Shucksmith 1981, 57) where all four district councils discounted some of those on their waiting list as not being in need. Until recent legislation in Scotland made it illegal (Tenants' Rights Etc.,(Scotland) Act 1980, section 26), it was common for non-residents to be excluded (e.g. Ettrick and Lauderdale, Orkney and many others), for owner-occupiers to be debarred (e.g. Sutherland and Orkney) and even single persons under forty were ineligible in Orkney (SDD 1979, 66). Even where such groups are now eligible to join a waiting list, they may effectively be denied access to council housing by exclusion from a priority list or by the operation of the points system. Households debarred in these ways are considerably disadvantaged and their housing opportunities significantly diminished. The groups most affected appear to be young households and other groups in private rented accommodation.

At the same time, need is also underestimated in the smaller villages because of the many factors which discourage individuals in rural areas from expressing their felt needs by joining a council waiting list. In many rural parishes there are no council houses at all, and as a result there are no waiting lists for those parishes: people are unlikely to queue for houses which do not exist. Even if they wish to, it may not be possible: a resident of Achmore, in Lewis, was encouraged by the local councillor to join the waiting list for council houses in his village although none existed at that time, but was told in the housing department of the Comhairle nan Eilean that no such waiting list existed (comment at public meeting). Similar experiences have been reported by Inverkeilor and District Community Council (personal communication). In Kinlochewe, in Ross and Cromarty, villagers have reported a desperate need for housing for young people who will otherwise leave the village, and complain that the housing authority insists there is no demand for housing (*West Highland Free Press*, 6 July 1984). Such reports are anecdotal, but they are so widespread and consistent that their substance is now generally accepted.

The problem is not confined to villages with no council housing. Even when there are a few council houses in a village, most of the local population are well aware that no vacancies are likely in the foreseeable future. Only when it becomes known that a house is to become vacant will many people express an interest in it. A recent survey of mobile home dwellers in Skye found that, while 78% wished to leave their temporary accommodation, only 28% had joined the council waiting list. 'It was clear when speaking with the residents that a number were not on the waiting list simply because they believed that they had no hope of getting a house' (Harbison 1983, 5). At that time there were 458 applicants on the waiting list in a district with a total council stock of 736 houses. Gordon District Council, which favours dispersal of population, has encountered this hidden need in more favourable circumstances. It has provided council houses, including sheltered units, where they were not strictly justified by waiting list figures, overcoming initial fears that units would remain unlet: as the schemes materialised, more and more people applied for them, having thought at first that they had so little chance of being housed that it would not be worth registering an application (personal communication). This appears to be a common pattern encountered in rural areas when rented housing is provided, and is increasingly recognised by both housing associations and local authorities (Highland Perthshire Community Councils Forum 1987).

Often the expression of need may be diverted, as individuals join waiting lists for council houses in the larger settlements, not because they prefer to leave their villages but because the chance of a vacancy occurring is so much higher in a town. Banff and Buchan District Council (1981) discovered this phenomenon through a survey of its waiting list applicants. A large number of the applicants for council housing in Mintlaw, a small agricultural town, were found to have come from the Deer area where housing was unavailable. Although the waiting list appeared to show a need in Mintlaw, and none in Deer, the survey revealed that the applicants' preference would have been to stay in the Deer area.

The existence of hidden needs, for all the reasons described, ensures that the extent of rural needs is seriously underestimated by waiting lists. It follows also that any study of inequitable access to housing which rests on an analysis of council waiting lists (expressed need) can only be partial. The analysis of council waiting lists is a valuable part, but only a part, of the study of rural housing advantage and disadvantage. It will be essential to place this within a broader frame of reference, which also considers both public and private sector allocation mechanisms and the housing circumstances of those whose needs remain hidden for one reason or another. Such an attempt to identify the pattern of rural housing disadvantage and advantage is made in Chapter 5.

Local needs revisited

In Chapter 3, however, the existence of a concept of 'local needs' in many structure and local plans was discussed, and it was noted that the use of this term in a vague and imprecise way appeared to relate less to the social concept of need, as reviewed in this chapter, and more to the politically acceptable rate of housing development in the area.

But even if the concept of need is elaborated to allow a more precise and operational definition of need, and then combined with some geographical boundary to arrive at a workable concept of local needs in the social sense, the question which eventually arises is whether local claims to rural housing should indeed have any special priority. As Rogers (1985a, 375) has noted, local needs have been accepted largely uncritically as a suitable basis for housing policy when the justification is by no means clear. The arguments for and against are discussed in detail by Rogers, and it will be useful to summarise these briefly at this point.

Rogers proposes five main arguments in support of 'locals only' policies. In the first place, local residence may act as a proxy for welfare objectives in housing policy since local people in rural areas are taken to be generally poorer and more deprived. This is clearly an over-simplification: while manual workers and other low paid groups may be highly disadvantaged, rural areas also contain many very wealthy people, and a policy which helps locals as a proxy for the poor is a rather blunt instrument. A second argument dismissed by Rogers is that outsiders have no serious grounds for being housed in the countryside.

A more popularly held view is that housing for local people contributes directly to local employment generation and support, and to the maintenance of services. As Rogers argues, this argument is popular 'with existing employers, particularly landowners who can combine a "locals only" policy with a general policy of restraint' and with others seeking to expand employment opportunities in the area. But, he continues:

> While this argument appears on the surface to provide a fairly strong and practical justification for local needs policies, it contains a contradiction in so far as localness can thus be justified in the case of those who move from outside the area to work within it. This type of local claim is, therefore, in direct conflict with those arguments which appeal to the idea of the locally-born person, beleaguered in the housing market by the influx of outsiders.
>
> (Rogers 1985a, 376)

A fourth argument, popular with local councillors, is that local people have a moral right to priority over outsiders, since local authorities are elected to reflect the interests of their electorate. A fifth view is a pragmatic acknowledgement that the label of 'local needs', whatever its limitations, is politically very powerful in that it has genuine popular appeal and demands media attention for rural housing issues. Thus,

> The amalgam of rural deprivation, grass roots politics and practical compromise between development and control which is represented by the label 'local need' provides a way of getting rural housing issues into the public and political arena.
>
> (Rogers 1985a, 376)

Against these five arguments in favour of local needs policies, Rogers presents five counter-arguments. The first two of these question the legality of such policies and the definition of local, respectively. The third set of arguments claims that these policies are ineffective or even counter-productive in that local people are disadvantaged rather than helped by them; this anticipates some of the discussion in relation to the case study in Chapter 6. A fourth argument is that a successful local needs policy would, by excluding outsiders, lead to stagnation and social atrophy. Finally, it has been argued that when taken to its extreme (as in Jersey and Guernsey) a locals only policy becomes both unjust and immoral.

Rogers concludes that there are so many doubts and uncertainties about the concept of local need that it provides a very poor foundation on which to build policy.

The use of local residence as a proxy for housing disadvantage, in particular, was criticised as an oversimplification, and in the remainder of this chapter and in Chapter 5 an attempt is made to identify more precisely the groups which are disadvantaged, both in terms of access to housing for consumption and in terms of access to the domestic property market and its associated accumulative potential. In Chapter 2, the desirability of constructing a Weberian typology of rural housing classes was accepted, and the final section of this chapter reviews the various typologies of social groups in the countryside which have been proposed, and suggests the main criteria which might be used in the identification of housing status groups and housing classes. A typology is then proposed to inform the discussion of rural housing advantage and disadvantage in Chapter 5.

Rural housing classes and housing status groups

Research into rural housing has often employed crude dichotomies such as local/newcomer and low/high income, often with an implicit assumption of correspondence between locals and low income groups, as illustrated in the justifications proposed for local needs policies. A similar tendency characterises many public policy statements, and structure plans in particular (Rogers 1985a).

Yet more sophisticated typologies of social groups in relation to their rural housing opportunities have been proposed, most notably by Pahl (1966) and by Ambrose (1974), which afford a greater insight into the social and economic consequences of public policy. Further, Saunders (1980) has outlined the skeleton of a typology of Weberian housing classes which places at its centre the means of access to the domestic property market, through which individuals may pursue the accumulative potential of property ownership and, additionally, the consumption of housing. Such a typology, applied to rural housing, might be expected to illuminate the more fundamental class basis of conflicting policy objectives, policy formation and distributional outcomes.

Saunders (1980, 93–5), on the basis of a reassessment of the Weberian approach, argues that because of the accumulative nature of their tenure, owner-occupiers 'constitute a "middle" property class whose interests may align with those of either the "positively privileged" class of housing suppliers, or the "negatively privileged" class of non-owners, according to the nature of the issue'. As noted in Chapter 2, this model represents an advance on Rex and Moore's conception of housing classes

> in that it denies the relevance of the ecological factor (ie. spatial segregation), it rejects an arbitrary pluralism in favour of a trichotomous model of housing classes, it emphasises the process of real wealth accumulation as a necessary criterion of analysis, it rests

> on an objective assessment of class interests rather than on any assumption of shared values, and it includes the suppliers of housing as one class in the property class system.
>
> (Saunders 1980, 94)

Saunders then proceeds to elaborate his model in a way which offers considerable potential insight in a rural context, disaggregating each class into further 'distinct strata'.

The first class is 'private capital engaged in the supply of housing or in the provision of facilities and services necessary to its supply and distribution' and the interests of this class lie in profit maximisation. Saunders identifies the major strata within this class as:

1. Finance capital: engaged in lending money for house purchase, housebuilding, and house improvement.
2. Industrial capital: engaged in house building.
3. Commercial capital: estate agents, solicitors and surveyors engaged in the market exchange of housing.
4. Landed capital: both landowners and private landlords.

Sometimes these strata will find their interests conflict, as finance capital seeks high interest rates when it lends to industrial capital for housebuilding, for example, but in general these strata share a common interest in the returns to housing as a commodity.

Specifically in the context of rural land use controls, one would expect the interests of these various strata to conflict. Housebuilders will clearly want land to be released for housebuilding in sufficient quantities to meet their demand predictions (assuming prices which maximise their profits) and at a steady supply which allows them to maintain a stable level of operation. Builders' organisations have therefore been highly critical of restrictive planning policies, which by raising both house prices and land prices result in little increase in builders' profit margins, and instead tend to reduce builders' scale of operations and increase uncertainties, delays and market fluctuations:

> The speculative builder has long been a *bête noire* for the shire counties, a threat to rural preservation, someone to be kept at bay at all costs. Housing developers regard planning constraints as a barrier to be overcome before their own plans can be brought to fruition. If it were not for these restrictions housebuilders could build wherever they liked. They see planning as restrictive and negative. It also makes it difficult for them to anticipate when land will be available, and it creates uncertainties during the interval between land purchase and construction.
>
> (Herington 1984, 133)

Neither, however, do builders favour an absence of planning controls, calling instead for 'proper regional planning' which would ensure a coordinated and highly regulated release of building land (HBF 1987). White (1986) and Barlow (1988), among others, have pointed to the crucial significance of developers' land-banking. While most housebuilders are thought to use their land banks primarily to ensure production turnover, while incidentally enjoying profits from land price increases, Rydin (1984) has suggested that some firms, particularly the larger ones, ' will favour a situation which allows *them* to constrain the supply of development land, thus meeting their twin objectives of general restraint in areas of high demand with the grant of planning permission for land in their ownership' and so conferring speculative gains on them from land hoarding.

From work on urban gentrification (Williams 1976), however, it appears that exchange professionals and building societies (commercial and finance capital) have an interest in encouraging a faster turnover in the domestic property market and an escalation of house prices so as to increase their commissions and interest respectively. These strata may therefore be thought to favour restrictions on new housebuilding in rural areas, to the extent that this causes house prices to increase and attracts more wealthy purchasers who are more secure mortgagees.

Landed capital's interests are more ambiguous. As Hall *et al.* (1973) have argued, the owners of land designated for development will gain and other landowners lose as a result of planning restrictions. Many may also be private landlords who will gain through the increased asset values of their domestic properties and from higher rents, whether from permanent tenants (whose rents although regulated are regularly reviewed) or increasingly from holidaymakers (whose rents are not controlled). Beyond this, Newby (1980, 239) has argued that landowners have a vested interest in preserving the status quo by excluding new developments which might upset the 'traditional rural way of life' and thus threaten their social and political supremacy. However, this may have changed as a result of the fall in farmland values during 1985–7 and the expectation of further decline as agricultural output prices fall. Thus, the CLA welcomed the proposal to 'free land for much needed housing in rural areas' (Guardian 10 February 1987) because this might allow their members to maintain their returns from the land.

On balance, then, the interests of property suppliers are likely to be divided in relation to the construction of new housing in the countryside. Where their profits depend upon the supply of new housing, as for housebuilders and the owner of the land in question, their interest would be served by releasing land for new housebuilding; however, other elements of capital will gain most by the protection of the rural environment and the consequent increase in the price of existing housing.

Of course, landowners' motives for holding land or selling land, and thus contributing to the residential development process, may be affected by personal and non-pecuniary considerations as much as by financial or investment criteria (Munton and Goodchild 1985). Among the factors influencing his decision to release land will be the landowner's material interests and social relations, covering kinship, social status and class. These factors are discussed further in Chapters 7 and 8, in relation to recent policy initiatives. It will be argued, in particular, that these factors will lead farmers and landowners to vary significantly in their willingness to release land for low-cost housing.

In contrast, the interests of owner-occupiers in relation to new housebuilding in rural areas are quite unambiguous, and this is the principal justification for the adoption of this typology. Owner-occupiers gain by restrictions on the supply of new house construction which tend to inflate the exchange value of their own houses, and this is true whether they are mortgagees or outright owners, locals or newcomers. Newcomers are often held responsible for opposition to new housing, through the

> so-called 'drawbridge' effect whereby the migration of higher status groups is followed by strong opposition to further growth. The pioneers express opposition outside and inside the formal political process. The more articulate middle class secure places on the local council and fight for conservation or low growth planning strategies, helping to reinforce social segregation.
>
> (Herington 1984, 148)

As Bell (1987, 23) has noted, 'the NIMBY syndrome (not in my back yard) is well known; and "last one into the village runs the preservation society" is but a jocular representation of reality'. But is this opposition to growth a defence of lifestyle or a reflection of domestic property class? Herington summarises Kramer and Young's (1978) explanation of opposition to growth in terms of territorial defence of lifestyle:

> Those who place a premium on their environment and enjoy a high standard of privacy and social exclusivity naturally seek to defend their lifestyle against the advancing urban tide.

but he also notes that:

> House prices have risen steeply under the combined effects of rising income, rising housing demand and shortage of supply. This might account for the vociferous opposition to further land for housing in landscapes of little scenic value.
>
> (Herington 1984, 148)

Although there is no reason to expect domestic property classes to share a common class consciousness, or to organise on a class basis, nevertheless it is revealing to consider the composition of local amenity societies to see if they are in fact dominated by newcomers in defence of their status. Buller and Lowe (1982) examined the interests and values of the Suffolk Preservation Society through a survey in 1977, with a view to testing the stereotyping of rural preservationists 'into two mutually exclusive groups – the old and the new': composed on the one hand 'of the old squirearchy whose desire for a preserved countryside goes hand in hand with a predilection for a preserved social order', and on the other of 'nouveaux rustics, city folk newly arrived in the countryside, who having found their ideal spot are reluctant to see any changes which might affect their chosen rural haven' (Buller and Lowe 1982, 24-5). In fact, both of these stereotypes are rejected, since 'most of the committee members are part of the local professional or business classes' and 'established Suffolk residents' of at least twenty years. 'Though not the traditional county establishment, the majority of members are drawn from the older and better off Suffolk residents. All are home owners' (p. 25). Buller and Lowe demonstrate that although the Society seeks to escape the charge that they are well-off people selfishly protecting their property interests and privileged environments, the Society nevertheless 'consistently opposes housing developments which seem out of character or scale with existing settlements' so that 'many fewer houses have been built and those that have been are up-market' (p. 37). This leads the authors to the conclusion that 'the Society's basic concern with the character of Suffolk villages is with property not people' (p. 38). In addition to long-standing societies such as the Suffolk Preservation Society and the Friends of the Lake District, Herington (1984, 152) notes that ephemeral 'locality-based pressure groups usually emerge when specific development proposals threaten the property values of local residents', and this illustrates further the propensity of home owners to resist new housing in defence of their property values.

The empirical findings of Buller and Lowe suggest that attempts to explain opposition to new housing simply according to newcomer/local dichotomies may be misleading: the evidence tends to suggest that opposition arises at least partly from the vested interests of home owners in protecting the potential of their properties for capital accumulation, quite apart from any common interest in preserving their lifestyles. This supports Saunders' argument that owner-occupiers must be seen as a distinct domestic property class because of their market interest in the accumulative potential of their homes, quite apart from their interest as consumers, at least in a rural context. At the same time, it is hard to distinguish this class orientation empirically from status groups organising to protect their privileged lifestyles, and this may therefore be a useful subject for further research. Short *et al.* have undertaken some work of this type with reference to Berkshire, characterising resident groups as either 'stoppers' or 'getters'. Their conclusions illustrate the intertwining of material and status concerns:

> The concerns of stoppers reflect both environmental and economic calculations, and also specific social valuations... There is a hard core of material interest underneath the environmental concern, relating to the impact of development upon house prices. This calculation also relates to the social composition of an existing neighbourhood and any new development. Many households have been attracted to certain villages... because of their exclusivity. Particularly in the rural areas there is a powerful ideology which sees in a village location the hope of restoring a moral arcadia away from the anonymity of mass urban society... In summary, the voice of the Stopper is the voice of middle-class, middle-aged, owner-occupiers seeking to protect their physical and social environments.
>
> (Short *et al.* 1987, 36–7)

Anti-development interests may therefore be viewed from this perspective as essentially domestic propertied interests.

The third class which Saunders identifies from a Weberian perspective is the non-owners of domestic property, who therefore stand to gain nothing from increased exchange values. Saunders (1980, 95) sub-divides this class further into council tenants, private tenants and the homeless. The common interest of all these is in increasing the size of the rented stock and in reducing rents and house prices. They are therefore likely to gain unambiguously from increases in the volume of housebuilding in rural areas and from the relaxation of planning controls. The most direct benefits to these households are likely to arise from the construction of rented housing, since additional owner-occupied housing is allocated by the market mainly to wealthier groups, whereas additional rented housing is likely to be built by the public sector and allocated according to need. However, as the next chapter demonstrates, even public sector allocations create patterns of advantage and disadvantage. Of course, some non-owners will benefit from increased speculative housebuilding in rural areas since this will tend to dampen price increases and so make home ownership more accessible than under severe planning restraint: however, they will be only a small minority. Particularly for those who are favoured by council allocation criteria, an increase in the volume of council building (and the provision of other forms of social housing) will be most significant. Households whose needs are given less priority by council landlords will also find their position improved by initiatives which seek to encourage low-cost home ownership or an increase in private tenancies.

Again, there is the possibility of conflicts of interest between different strata within this class, notably over the issue of council house sales. Indeed, it could be argued that council house tenants in rural areas with a right to buy are more properly members of the class of owner-occupiers, since they have gained the right to acquire an appreciating asset at a substantial discount, and may therefore now have a vested interest in an increase in the exchange value of domestic property.

In summary, therefore, a Weberian typology of domestic property classes in the rural context might be elaborated as follows:

1. Suppliers: (a) suppliers of new houses (builders, landowners);
   (b) those whose profits derive from house exchange.
2. Owner-occupiers
3. Non-owners: (a) council tenants;
   (b) private tenants;
   (c) homeless households.

Saunders himself has offered an auto-critique of his own Weberian approach which led him to reject it (Saunders 1980, 96–8); however, Pratt (1982) has re-examined this approach and concluded that Saunders' self-criticisms are weakly sustained and that tenurial divisions can form the basis of class formations in the Weberian sense. Without reviewing these arguments in detail, it is worthwhile to make the point that in a rural context Saunders' self-criticisms seem even less substantial, and that the concept of domestic property classes may usefully be applied to rural housing notwithstanding.

Of Saunders' criticisms of his own domestic property class analysis, three are most serious. These are the historical contingency of the analysis; the absence of conflict between all property classes; and the uncertain correspondence between domestic property classes and acquisition classes. Saunders (1980, 98) regards as 'the most serious weakness of this whole approach that in the final analysis it is empirical and descriptive, dependent upon the continuation of certain specific conditions which are external to the analysis itself'. If these historical conditions – namely real house price increases, negative real rates of interest and tax subsidies – altered, then Saunders argues that owner-occupation would no longer be a source of capital accumulation, and the basis of his typology would collapse. However, in the context of rural housing two counter arguments apply: in the first place, the conflict over land release for housebuilding (which is the primary determinant of real rural house price increases) is endemic and not a temporary phenomenon, given the existence of a development control mechanism; and second, and related to this, the conditions which have created real rural house price increases are not external to the analysis, since the rate of land release for housebuilding may be strongly influenced by the opposing classes, as we have seen. These arguments suggest that in the rural context the domestic property class analysis does not rely on unduly specific external conditions.

Saunders' second major concern was that, although he was able to establish class conflicts ('relations of exploitation') between housing suppliers and home owners and between housing suppliers and non-owners, he could not demonstrate a necessarily antagonistic relationship between home owners and non-owners. Yet in the rural context the conflict over land release for housebuilding, and therefore over house prices and rents, necessarily places home owners and non-owners in material opposition to one another. Restrictions on development increase the exchange value of home owners' property while increasing rent levels, reducing the supply of homes to rent, and further limiting accessibility to owner-occupation. Thus, Edel (1982, 219) has noted that conflict occurs locally, where 'owner-occupier may be set against tenant over issues of development vs growth control, and other environmental and regulatory questions': however, this is no local aberration but an inherent feature of rural property markets. Further, as Pratt (1982, 490) has argued, when non-owners are excluded from their neighbourhoods by the gentrification process, the newcomers purchasing homes in the area are 'exploiting them in so far as they are dominating the spatial location through their greater market power'.

Saunders' third fundamental criticism was that the relationship between domestic property classes and acquisition classes was unclear, and yet this may not be a serious weakness. Pratt (1982, 493) suggests 'that the relations between domestic property classes...and acquisition classes are open, empirical ones – not ones that a priori invalidate the domestic property class analysis'. If this suggestion is accepted, then it is clear empirically that in rural areas there is a close relationship between acquisition classes and domestic property classes. Access to home ownership depends upon either wealth or access to credit which, in turn, depends on size and predictability of income. Owner-occupiers in rural areas are predominantly professional and non-manual workers (Rogers 1983) or people retired from these acquisition classes. Nevertheless, access to home ownership and associated capital accumulation is not determined solely by labour market power: other relevant factors, as we shall see, include life-cycle stage and inherited wealth (Dunn *et al.* 1981). For this reason domestic property classes and acquisition classes, although closely related, may be seen as analytically distinct.

Therefore, the concept of domestic property classes appears more robust, at least in the rural context, than claimed by Saunders. How does it relate, however, to previous attempts to define social groups in the countryside? Pahl (1966) and Ambrose (1974) have each attempted to construct typologies of social groups in rural England based largely on housing status, and Dunn *et al.* (1981) have constructed rural housing profiles which reveal the main social groups competing for housing in rural areas.

Previous attempts at social typologies

Pahl stressed the heterogeneity of the village population, and urged planners to eschew 'sentimental olde-worlde notions of community' and the 'vague romantic flub-dub' of the Scott Report. His typology was not based on any actual village, 'but arises out of an analysis of field studies and surveys undertaken in south-east England in the 1960s by a number of research workers and planning departments' (Pahl 1966).

Pahl regarded his proposed social groups as a preliminary framework for future research, and acknowledged that his assertions would be hard to test empirically. The main criteria he used were housing tenure and ownership of capital/income, although some emphasis is also placed on the motivation for living in a village and how this relates to the individual's perception of the village – Pahl's 'village in the mind'. His focus is exclusively on housing consumption rather than on accumulative potential, and his typology therefore rests on consumption cleavages, not domestic property classes. Dunn *et al.* (1981) largely accept, and elaborate upon, Pahl's generalised framework, but suggest that four further groups be added: second home owners, mobile home dwellers, tenants of winter lets, and members of the armed forces.

Ambrose's discussion of rural housing arises out of his analysis of social change in the Sussex village of Ringmer during the last century. This leads him to question the present organisation of housing provision on the grounds that it is spatially and socially divisive, regressive in its distributional impact, and fails to satisfy housing need. He claims 'to discern rough groupings of households', again 'based on general observation rather than empirical data' (Ambrose 1974, 201).

His typology is similar to Pahl's insofar as it stresses the ownership of capital/income in relation to access to different housing tenures, and is focused on consumption cleavages rather than accumulative potential. However it adds the element of car ownership, on the basis that being without a car restricts one's choice of housing location. Ambrose argues that without a car households are discouraged from living in villages. Further, the wealthiest tend not to live in villages because two-car households are able to live in seclusion outside villages, and because of a shortage of four-bedroomed owner-occupied houses. Similarly, a dearth of privately rented houses (other than tied cottages) excludes those who cannot afford to purchase and do not qualify for a council house. This sorting process tends to leave two groups (those who can afford a semi-detached house, and those who cannot afford to purchase but whose situation enables them to gain council housing) alone in the village, distinguished from one another primarily by housing tenure, which then becomes the axis of social polarisation.

It was suggested earlier that the most fundamental distinction which must be made in identifying housing classes is between those who can buy (and so set foot on the ladder of potential capital accumulation) and those who must rent. The typologies of Pahl and Ambrose suggest further cleavages in relation to consumption, on the basis of car ownership and one's motives for seeking rural housing, in particular. Life-cycle is another factor which might be incorporated, given its significance in urban housing models (Short 1982). While these might form the main dimensions of consumption cleavages in rural housing, it is quite possible that their application to any particular area would require further elaboration and refinement.

Several of these dimensions are included in a model of rural housing opportunity developed by Dunn *et al.* (1981, 138).

> Stage in the family life-cycle and socio-economic status provide the essential starting points for this model. For most households these two provide the key to household income and thence to the central aspect of tenure, though for a growing minority inherited wealth in the form of property will provide an equally important starting point.

The authors argue that housing tenure and car ownership are the principal determinants of quality of housing (its type, condition, amenities and age), quality of life (education, health, services and recreation) and employment (type and access). These, in turn, determine the individual's socio-economic status, so establishing a cycle of deprivation and advantage, with housing tenure at its core. Their model only makes passing reference to the accumulative potential of rural home ownership, however, so that their work, like that of Pahl and Ambrose, cannot be seen as an attempt at class analysis, although its emphasis on differential access to housing is closely related to Weberian concerns.

Dunn *et al.* attempt to go beyond this, however, by relating this model of rural housing opportunity to an empirically derived typology of rural housing environments, or rural housing 'profiles'. These rural housing profiles are then cross-tabulated with a large number of census variables relating to family life-cycle, socio-economic group, car ownership and housing tenure, in order to suggest groups which share similar housing circumstances – i.e. housing status groups rather than domestic property classes. Since this represents the only attempt thus far to identify social groups in relation to rural housing 'by rigorous statistical analysis', rather than by general observation, a fuller account of this work is now required.

The rural housing profiles were derived from a cluster analysis of 1971 census enumeration districts (EDs) in England. The clusters formed are therefore groups of similar EDs rather than housing status groups or housing classes composed of households or individuals. Nevertheless, Dunn *et al.* propose their cluster analysis in the context of a classification of rural households (1981, 79–90), and describe the work as 'a classification of rural households to give seven rural housing profiles' (p. 119). They conclude a discussion of the clusters of EDs with the claim that 'this has culminated in the recognition of the main social groups which are in competition for housing in rural areas' (p. 90).

The means by which such a typology of rural social groups emerges from a classification of EDs is far from clear. It appears that for each variable employed in the cluster analysis the mean value for EDs within that cluster was compared with the national mean. For example, one cluster, labelled 'agricultural: farmers', included mainly upland EDs from north and west England. Within these EDs the number of farmers and farmworkers, the proportion of households in private rented unfurnished accommodation, and the population of pensionable age were all above the national average. This leads to the conclusion that:

> At a national level this profile represents one of the classic forms of rural housing, showing a familiar set of housing and related problems: an ageing population housed in poorly equipped housing stock with the attendant problems of poor service provision, isolation and few employment alternatives to agriculture.
>
> Dunn *et al.* 1981

While not disagreeing with this conclusion, it must be questioned to what extent this picture of a distinct ageing social group is derived from the analysis of thirty census variables, as presented, and to what extent it is an *a priori* category consistent with (but an embellishment upon) the empirical information available from the cluster analysis. This is not to dispute the typology of social groups identified – indeed it concurs with other research findings. However, it does seem that their conclusions, like Pahl's and Ambrose's, rest as much on *a priori* categorisation and general observation as on the cluster analysis of EDs.

A further difficulty is that a method which compares profile means with national means in order to identify social groups will inevitably fail to identify the diversity of social groups within each cluster of EDs. For example, the cluster labelled 'Owner-occupiers: high status' shows a mean of 20.7 per cent of households with two cars, compared to a national average of 15.8 per cent. This might be taken to imply an absence of problems of accessibility and isolation, and yet 79.3 per cent of households do not have two cars. Indeed, 27 per cent have no car, and even though this figure is less than the national mean it indicates a substantial minority who should be included in any classification of rural households. It is notable that this cluster includes the metropolitan villages studied by Pahl, as Dunn *et al.* acknowledge. Pahl proposed eight social groups from his knowledge of such areas, but it is hard to see how these could all be identified through rural housing profiles. The danger of the method of rural housing profiles is that most of these groups will pass unrecognised.

Dunn *et al.* propose seven rural housing profiles, on the basis of their cluster analysis:

1. agricultural: farmworkers
2. agricultural: farmers
3. owner-occupiers: retired
4. transitional rural: accessible countryside strongly influenced by urban forces
5. owner-occupiers: high status
6. armed forces
7. local authority housing.

More localised analysis of EDs in the Cotswolds and South Oxfordshire suggested slightly different clusters, but again the main divide was between truly 'rural' housing profiles (the agricultural and the retirement areas) and areas demonstrating 'urbanising' characteristics. The latter set was sub-divided on tenure lines between areas of owner-occupiers and transitional rural on the one hand, and areas of local authority and armed forces housing on the other. In spatial terms, it is suggested that these profiles correspond to a core–periphery pattern, which emphasises the importance of the urbanising influence (p. 86). Despite the shortcomings of the rural housing profile approach, it is valuable insofar as it suggests the importance of migration and tenure related dimensions in understanding rural society. These aspects of their findings are confirmed by further cluster analyses undertaken by Harper (1987a).

By a crude ranking process, a substantive conclusion is reached about the incidence of rural housing advantage and disadvantage:

> The picture which emerges...emphasises the great divide in housing circumstances which exists between, on the one hand, the established and newly arrived professional groups in rural areas and, on the other, the rest of the rural population. The relatively poor position of private tenants, particularly those in tied accom-

> modation, is also made clear. By comparison with them the circumstances of local authority tenants are noticeably more favourable. Finally...the more urbanised rural areas...are at a clear overall advantage when compared with more remote countryside.
> (Dunn *et al.* 1981, 160)

For the purposes of this book, no attempt is made empirically to derive housing status groups through multivariate analysis, largely because no data on individual households are available which would allow this (although see Shucksmith [1990, forthcoming] for an application of such an approach in the atypical circumstances of crofting areas]. Housing status groups and domestic property classes are composed of individuals or households, and these are the appropriate levels of analysis for such a categorisation, not EDs. Without access to individual census returns, or detailed surveys of households, housing status groups and domestic property classes must be proposed on the basis of non-survey methods such as participant observation, or suggested *a priori* through a process of deductive reasoning informed by previous research and general observation. The latter approach has been seen to be that of Pahl (1966) and Ambrose (1974), and to a large extent it is also the basis of the groups identified by Dunn *et al.* (1981), despite their claims to have derived these categories empirically.

The typologies offered by Pahl, Ambrose and Dunn *et al.* return us briefly to our methodological discussions, in order to assess their utility in relation to the objectives of this book. While perhaps less relevant to the theme of land use conflicts than is the concept of domestic property classes, typologies of social groups in relation to their differential access to housing consumption are nevertheless relevant to our concern with distributional outcomes. It follows that there may be some value in elaborating consumption cleavages, in addition to focusing on domestic property classes and their accumulative potential. The necessity of analysing consumption cleavages also appears to be the final position which Saunders adopts, when he argues that housing tenure

> is analytically distinct from the question of class; it is neither the basis of class formations (as in the neo-Weberian tradition) nor the expression of them (as in the neo-Marxist tradition), but is rather the single most pertinent factor in the determination of consumption sector cleavages. Because such cleavages are in principle no less important than class divisions in understanding contemporary social stratification, and because housing plays such a key role in affecting life chances, in expressing social identity and (by virtue of the capital gains accruing to owner occupiers) in modifying patterns of resource distribution and economic inequality, it follows that the question of home owner-

ship must remain as central to the analysis of social divisions and political conflicts.

(Saunders 1984, 207).

Consumption cleavages

If housing classes are identified in relation to consumption, the criteria most germane to a typology are those which most strongly influence an individual's or a household's chances of gaining access to housing. Ownership of capital and level of income are the most fundamental determinants. Those who own capital, usually in the form of a house, are already set upon the ladder of escalating property prices: this gives them both the ability to trade their capital for owner-occupation, the collateral for borrowing, and the benefit of any rise in property values (which from this perspective may be seen as enabling them to trade houses and gain access to even higher levels of consumption). This is not to say that all owner-occupiers are well off – indeed the proportion of home owners in serious financial difficulty is increasing – but in the main their housing consumption possibilities are greater. A high income also confers clear advantages, both in terms of ability to repay a mortgage, to qualify for a certain size of mortgage, and to meet maintenance and other outgoings associated with house purchase. These circumstances can be contrasted with those of a household lacking capital and receiving a low, irregular income, whose housing consumption options may be much more constrained: almost certainly this household will have to rely on rented accommodation, and this may have to be private unless other characteristics of the household are such as to gain it a sufficiently high priority on a council waiting list.

The existing tenure of a household is also a factor in its future housing consumption possibilities. Once allocated a council house, for example, a household has a wider set of opportunities, including the right to remain in that house as a secure tenant, the right to buy it at a discount, the right to transfer to another landlord, and access to the transfer list. The wider access to home ownership enjoyed by an existing owner-occupier may be offset in the case of impecunious home owners by the points penalties they face in any attempt to transfer, or revert, to council housing. Similar obstacles may face those in tied housing in Scotland or in winter lets.

Income/capital and existing housing tenure are likely to be the main factors determining who can buy and who must rent, which has already been proposed as the major distinction between domestic property classes, as well as between housing consumption 'classes'. But several other factors will also influence access to housing consumption.

The household's motives for living in the area are particularly relevant, since there may be considerable differences in the housing opportunities open to retirement migrants, 'pastoral migrants' (Forsythe 1980), second home buyers, active commuters, reluctant commuters (Pahl 1966), indigenous people in local employment and other groups. It is likely, for instance, that households tied to a particular workplace within the rural area will have a more restricted pool of dwellings from which to draw, compared to groups seeking no particular locality but a generally attractive rural environment. The lack of a car would emphasise this difference. This is quite apart from the suggestion, already reviewed, that such groups may also conflict in their attitudes towards proposals for new development and council house provision (Newby *et al.* 1978). Divisions defined by motives for living in the countryside may correspond to some extent with wealth and income divisions. There is a presumption that second home buyers, and indeed most migrants from urban areas, are wealthy relative to manual workers in traditional rural occupations, for example. Nevertheless, it will be helpful to distinguish analytically between different categories of local population, particularly according to wealth and income.

As regards housing consumption opportunities, life-cycle factors are also pertinent. Studies of council house waiting lists (Shucksmith 1981; Clark 1982; Phillips and Williams 1982) have shown that considerable expression of need comes from young couples, perhaps waiting to become married or already married and living with parents, unable to accumulate the necessary points to gain access to council housing until they have one or two children. Similarly, another major source of need for council housing comes from the many elderly people in rural society. The shortage of one- and two-bedroomed houses, both in the public and private sectors, may well contribute to this difficulty of housing access.

Taking account of these criteria, a typology which reflects differential housing consumption opportunities may be proposed as follows:

A. Low income, low wealth

1. Young households, often young couples and single persons from the immediate area, unable to buy into home ownership and denied access to council housing through insufficient points. This group tends to have to seek private rented accommodation, winter lets and mobile homes, or share with their parents.

2. Other tenants of private rented housing and tied accommodation, often in low paid, traditional rural occupations. Members of this group may be trapped in inadequate housing, with little prospect of being allocated a council house or of finding another private tenancy.

3. Pensioners, either retired from local employment or erstwhile

retirement migrants who now face difficulties due to increased living costs and accessibility problems. Members of this group could be in any tenure, and often seek greater support perhaps from a move to be nearer relatives or services, or from access to amenity or sheltered housing. While they enjoy the accumulative benefits of owner-occupation, they may also find difficulty in maintaining their home, and so its value.

4. Reluctant commuters, according to Pahl (1966) may have been forced to live at a distance from their urban workplace by an inability to purchase a suitable house in town. Nevertheless they will gain from the increase in their home's value, and this then will widen their options.

5. Local authority tenants are a small but relatively fortunate class compared to these other groups. However, their rents have recently risen considerably, and they may have had to leave their preferred locality in order to secure a council house in a nearby town. They will be eligible to buy their houses at a discount, and so gain access to the accumulative potential of owner-occupation.

B. More prosperous groups

6. Indigenous owner-occupiers, tradesmen, farmers and landowners have a considerable choice within the owner-occupied sector. They also share a proclivity for attempting to prevent further housing development, thus inflating their own property values and maintaining a low-wage economy (Newby *et al.* 1978).

7. Retirement migrants and 'pastoral migrants' will have ready capital available from the sale of their previous home, probably in an urban area, and will therefore enjoy a wide choice in the owner-occupied sector. They will also want to protect their property values, and the rural environment, by opposing new development. However, they are less likely to be tied to any particular locality.

8. Holiday home buyers will have a similar desire to preserve the rural environment and their property's value, but are likely to have less capital at their disposal for the purchase of rural housing (having a main home to finance as well). They may therefore compete to some extent with low income, low wealth groups for the cheaper end of the owner-occupied market.

9. Commuters (other than reluctant commuters) choose to live in a rural environment, rather than in a particular community, and again have a vested interest in restricting development and so increasing the value of their property. However, they are constrained by the need to live at an

acceptable distance from their urban workplaces. In common with groups 6 and 7, they will have a wide choice within the owner-occupied sector.

An attempt has been made to relate this typology of differential housing consumption opportunities (referred to henceforth as 'housing consumption classes') to the earlier typology of domestic property classes, although the two are analytically distinct[1], by including comments on each type's interest in relation to the release of land for housebuilding. It should be noted, however, that housing suppliers are excluded from the typology of housing consumption classes.

This proposed typology is not intended to be rigid or universal. It would clearly require modification in its application to a crofting area, for example. Nevertheless, it is argued that the criteria adopted for distinguishing housing classes – income/wealth, present tenure, life-cycle status and motives for residing in the rural area – will remain relevant and form the basis of any amended typology. The validity of this argument can best be judged in relation to the review of rural housing problems which follows (Chapter 5) and in the context of the case study.

Conclusion

This chapter has attempted to provide a framework for pursuing the theme of differential access to rural housing and the distributional outcomes of policy. The concept of need was explored in relation to inequity and disadvantage, and several conceptions of need were related to the measures commonly employed. It was concluded that an examination of rural housing advantage and disadvantage could not rely solely on the evidence of council waiting lists, but should also consider evidence of hidden need.

An examination of the validity of local claims to rural housing, implicit in local needs policies, led on to a discussion of rural housing classes. It was argued that Saunders' concept of domestic property classes was particularly relevant to the analysis of rural housing markets, given the inherent conflict between owner-occupiers and non-owners over the release of land for housebuilding. A typology of domestic property classes in the rural context was therefore proposed. However, earlier attempts to identify social groups in the countryside had not adopted this approach, but had instead attempted to identify housing status groups (focusing on housing market outcomes) or housing consumption classes (focusing on differential access to housing consumption). Since this book attempts to analyse differential access to rural housing as well as conflict over land release for rural housebuilding, it appeared relevant also to consider a typology of housing consumption classes. A typology which relied on criteria of income/ wealth, present tenure, life-cycle status and motives for rural residence was tentatively proposed, and this will form the basis of the analysis of rural housing advantage and disadvantage in Chapter 5.

Notes

1 Indeed, housing consumption classes are not truly classes in the Weberian sense at all (see Chapter 2). Nevertheless, this is a convenient shorthand term.

Chapter five

Access to rural housing: empirical evidence

Inequitable access to accommodation

The evidence of waiting lists

Analysis of waiting list information, and consideration of allocation systems, may offer some evidence of which groups are disadvantaged and which are relatively favoured. But the first point which must be made concerning waiting lists is that those who have joined one are themselves disadvantaged relative to most owner-occupiers, in relation both to housing consumption and to accumulative potential. This issue will be returned to later in this chapter, when discussing the shortage of council housing in rural areas. This section, however, reviews several studies of waiting lists in rural areas with the primary intention of gaining an impression, albeit partial, of which groups in rural society exhibit unsatisfied need for housing. Some insight into how council allocation policies modify the incidence of disadvantage may also be gained.

Most waiting lists include information about the applicant, their medical condition, and their present housing circumstances. The Scottish Development Department's (SDD) unpublished study of housing in rural Scotland (1979) examined the waiting lists of six case study areas. The most numerous groups applying for council housing in these areas were married couples with children, who were most likely to be allocated houses, but childless married couples and single persons together comprised over half of all applicants. Table 5.1 shows that their chances of being allocated a house were substantially less.

The study also found that most heads of applicant households were aged either under 30 or over 60, and that the more elderly applicants included a high proportion of single, widowed, divorced and separated heads of household. Again, allocation tended to favour certain ages of applicant disproportionately. Table 5.2 shows that applicants aged under

30 had a very much better chance of being offered a house than an elderly applicant, and it is suggested in the report that this is largely because of the general shortage of small, possibly specialised, two-apartment houses (SDD 1979, 73). Young, married applicants are most likely to be successful, especially if they have young children.

Table 5.1 Council house applications in rural Scotland, 1977

Type of household	*Applicants* (%)	*Allocations* (%)
Single persons	24	16
Married couple, no children	23	23
Married couple with children	29	44
Single parent family	6	8
Mixed adult family group	16	8

Source: SDD (1979, Tables 23, 27)

Table 5.2 Age of council house applicants in rural Scotland, 1977

Age of head of household	*Applicants* (%)	*Allocations* (%)
16–30	24	39
31–45	19	25
46–60	19	11
61+	30	15
not known	8	10

Source: SDD (1979, Tables 24,28).

Analysis of the tenure of housing applicants in rural Scotland revealed that almost half were living in private rented and tied accommodation. On average, a quarter were living in private rented accommodation and almost a further quarter lived in tied housing. The next most numerous tenures were households sharing with relatives, and owner-occupiers although some groups were locally more important, such as caravan dwellers in the Western Isles. Table 5.3 presents this information in detail.

Table 5.3 Tenure of housing applicants in rural Scotland, 1977 (percentages)

Owner-occupied	13
Local authority rented	8
Private rented	25
Tied housing	24
Sub-tenancies	3
Sharing with relatives	14
Caravans	6
Furnished rooms, etc.	4
Other	3

Source: calculated from SDD (1979, Tables 26 & 7A)

Most applicants were locals, even where waiting lists were open to non-residents, and the report comments that 'locals are particularly well catered for' (SDD 1971, 73).

Further information for one of the case study areas examined by the SDD (1979), Argyll and Bute, can be gleaned from that district council's housing plan 1980–5. That disaggregates the waiting list in 1979 into a priority list and a miscellaneous list, showing the applicant's principal reason for joining the list, and the size of house required. This information is summarised in Table 5.4. It can be seen that the single largest category of applicant is that of homeless persons who comprised 61 per cent of the priority list and 25 per cent of the full list. Applicants from other areas, mainly Strathclyde (SDD 1971), were almost as numerous, but the council placed these on its miscellaneous list, and only normally considered them for housing when a surplus existed. Most applications were for two and three apartment houses: the chronic sick and those living in 'below tolerable standard' (BTS) housing required mainly smaller houses, although half the homeless applicants required three apartment houses. This information confirms the disadvantage experienced by the elderly and infirm, and by tied tenants, but augments the SDD findings by suggesting the scale of homelessness.

Perhaps the major study of housing waiting lists in a rural context is that undertaken by Phillips and Williams (1982) in the South Hams district of Devon. Its evidence of the characteristics of waiting list applicants is remarkably similar in many respects. Again, most applicants are aged under 30 or over 60, and three-quarters of all families on the list consist of no more than two persons, as shown in Table 5.5.

Phillips and Williams also note that more than half those on the waiting list had no security of tenure, living for example with friends or relatives, in caravans, hotels or winter lets, or in tied accommodation. Large numbers lacked basic amenities, such as the exclusive use of a kitchen, bath or wc. A sizeable minority of applicants owned their own homes, and Phillips and Williams suggest that these are 'presumably the elderly' (1982, 90). In terms of their socio-economic characteristics, only 56 per cent are economically active, the remainder being unemployed or retired, and car ownership is low. Phillips and Williams conclude that 'the least well-off, the semi-skilled and unskilled manual, do figure disproportionately both as existing tenants and as applicants' (1982, 90).

Table 5.4 Argyll and Bute waiting list, June 1979

(a) Priority applicants (1,069)	*Total*	*Percentages*
Homeless	650	25
Overcrowded	69	3
In BTS housing	153	6
Key workers	51	2
Chronic sick	146	6
(b) Miscellaneous applicants (1,553)		
From outwith district	628	24
Adequately housed	398	15
Tied tenants	321	12
Engaged couples	142	5
Armed forces	64	2

Source: Argyll and Bute District Council (1979)

Table 5.5 Council house applicants' age and family size in South Hams, 1980 (percentages)

Age		*Family size*	
18–29	26	1–2	72
30–59	37	3–4	23
60+	36	5+	4

Source: Phillips and Williams (1982, 89)

Their study also examined the reasons put forward by applicants for their seeking housing. Three principal causes were put forward: the unsuitability of their present property or the expiry of a lease; changes in family circumstances such as childbirth, marriage or separation; and health reasons. In general, 'waiting list applicants tend to be those starting a family or those with immediate housing problems such as imminent homelessness' (Phillips and Williams 1982, 93).

The main sections of rural society identified by waiting lists as being in need of housing are therefore distinguished mainly by life-cycle stage, existing tenure and inability to buy. This accords with the typology of housing consumption classes proposed in Chapter 4, and it is clear that waiting lists include many of the first three classes proposed. In waiting lists, the only group of owner-occupiers which figures largely is the elderly, whose houses may be too big and too expensive for them to manage. Since these studies were done, another group of owner-occupiers may also have joined waiting lists in significant numbers (although this suggestion is unresearched as yet in a rural context): these are former council tenants who exercised their right to buy their homes, but now find themselves unable to maintain their mortgage repayments.

A much larger disadvantaged group is seen to be those tenants, young and old, who have no security of tenure. Either they are homeless, or homelessness is threatened. They live with relatives, in caravans or winter lets, in tied or other private rented accommodation. Some may have a good chance of gaining access to council housing, while others do not. This question of the distributional consequences of public sector housing allocation is considered next.

Councils have a statutory obligation to give priority in rehousing to homeless persons who have lived in the district for more than six months, and to those whose houses are to be demolished. In addition, in England and Wales (but not Scotland) councils must give priority to agricultural workers living in tied accommodation which is required for a replacement worker. These obligations do not, however, necessarily ensure housing for these groups. In many areas councils use wide discretion in determining whether homeless people are 'intentionally homeless', and therefore outside the protection of the Homeless Persons Acts.

For example, in Scotland a tenant of an agricultural tied cottage has no security of tenure, and no right to be rehoused by the council. Once evicted, he can apply for council housing under the Housing (Homeless Persons) (Scotland) Act 1977. Some councils, however, argue that a sacked farmworker is in the intentionally homeless category because he is to blame for his dismissal, and therefore that he is ineligible for housing (Hunter 1980). According to Shelter (1981, 27), 'some district councils have a very poor record, declaring over 40 per cent of their priority need applicants as intentionally homeless'.

Beyond those whom local authorities accept that they have a statutory obligation to house, priorities for rehousing vary from one district to another. In South Hams, Phillips and Williams found the next most favoured groups to be keyworkers and those with medical grounds for being housed: those without strong medical grounds 'could find themselves fairly low in the priority list for rehousing, despite the high number of points they may have been awarded' (Phillips and Williams 1982, 118). It was found that ordinary waiting list applicants had a much poorer chance of being rehoused during times of financial cutbacks than at times when the council could build in substantial numbers. This was because 'waiting list applicants are in a residual position in the allocation process' (Phillips and Williams 1982, 124), receiving only the houses which remain, if any, after the council has housed those groups which by statute must be given priority. This susceptibility to financial cutbacks is illustrated clearly in Table 5.6.

Table 5.6 Lettings in South Hams, 1976–80 (percentages)

Lettings to	1976–7	1977–8	1978–9	1979–80
Homeless	16.2	10.0	33.8	29.7
Key workers	2.4	1.1	3.4	3.7
Farm workers	1.6	0.5	2.9	4.1
Waiting list	79.8	88.4	59.8	62.5
Total numbers	124	557	204	219

Source: Phillips and Williams (1982,124)

Housing completions in South Hams have been cut back since the peak year of 1977–8, and it can be seen that the priority groups (homeless, key workers and agricultural workers) have each increased their share of lettings since then, leaving a significantly lower share for waiting list applicants. In absolute numbers, lettings to homeless people have changed little since 1977–8 (following a once for all increase in 1977–8, due to the Housing (Homeless Persons) Act 1976), the figures being 56, 69 and 65 for the last three years shown. And yet by 1978–9 those lettings represented about one third of the reduced total of all lettings. Thus, 'for local waiting list applicants, the statutory obligations to rehouse various other categories of people can mean a long wait to receive, eventually, inappropriate accommodation' (Phillips and Williams 1982, 147).

Of the general waiting list applicants, it has already been noted, on the basis of the evidence from rural Scotland that elderly applicants tend to be disadvantaged in council house allocation, largely because of the shortage of smaller houses and flats which would be suitable. And conversely, it was observed that young married couples had the best chance of being rehoused. These findings, together with a recognition of the residual position of those on the general waiting list, offer an insight into which groups in rural society exhibit unsatisfied housing need.

These patterns of advantage and disadvantage are not necessarily evidence of inequitable council house allocation policies, since it may well be accepted that the favoured groups, particularly the homeless, are those in greatest need. While accepting this, it seems that unsatisfied needs are also exhibited by many of the less favoured groups further down the waiting lists. If the most needy groups have the best chance of obtaining council housing, and therefore it is accepted that council houses are allocated broadly in accordance with need, the unsatisfied needs which remain are less the result of council house allocation policies, but rather are the result of insufficient council house provision. In this sense the resources allocated to housing are inadequate to meet needs, either because local authorities devote insufficient resources to council house building or because central government, through insufficient capital allocations and revenue subsidies, prevents councils from building in greater numbers. Whether or not more resources should be devoted to council housing in order to meet these unsatisfied needs is a complex question, requiring knowledge of the opportunity cost of those resources in terms perhaps of other unsatisfied needs, and requiring a value judgement to be made. The prescriptive sense of the word 'need' should not obscure the likelihood that global resources are insufficient to meet all needs, and that priorities have to be established.

However, council housing is only one of a number of housing possibilities for people seeking accommodation. Alternative tenures such as owner-occupation and private renting have their own patterns of advantage and disadvantage, and these will now be reviewed.

Obstacles in the private sector

The major barrier to becoming an owner-occupier is a financial one. In the rural areas of England and Wales in 1981, 62.5 per cent of households were owner-occupiers (DoE 1988b), and analysis of 1971 Census data shows that owner-occupation in rural England was 'very much the preserve of the professional and non-manual worker' (Rogers 1983, 114). In rural Scotland in 1971, a much smaller proportion, 37 per cent, of households were owner-occupied, and even this figure is misleading in that it includes households under crofting tenure (SDD 1979, 6). Nevertheless, the SDD's study reached similar conclusions, finding that, while prices had risen, 'average incomes in rural areas generally appear to have remained low and it is now only the more highly paid, often incomers, who are in a position to acquire much of the property available' (SDD 1979, iv).

The reasons for this disparity between high rural house prices and the low incomes of many rural households are several. Partly it arises from the relatively low wages paid in rural, manual occupations. Farm workers, tourism and catering workers, fishermen, hill farmers and crofters all receive lower than average incomes, as demonstrated by Thomas and Winyard (1979) and Mackay and Laing (1982, 105). The proportion of employees receiving low incomes also tends to be higher in rural and remote areas: the correlation between the incidence of low incomes and rurality, by county, was demonstrated in Shucksmith (1981).

Despite this incidence of low incomes, the evidence of several studies is that house prices remain high in rural areas (Dunn *et al.* 1981, 233–9). Partly this is because there are also many rural households receiving high incomes: McLaughlin (1985, 79) has demonstrated 'that those at the top of the rural league are considerably better off than their national equivalent whilst those at the bottom are much worse off'. Thus, the average gross weekly earnings for full-time males in rural areas were found by him, for non-manual workers, to be 32 per cent above the national average for non-manual workers, and for manual workers to be 4 per cent below the national average for manual workers. The full-time, male, non-manual worker in rural England averaged £215 per week, compared to £117 per week for a manual worker. This equates to the difference between a £33,500 and a £18,000 mortgage according to standard building society practice (three times annual salary). In addition to more affluent competitors for housing from within rural areas, there may also be competition from external sources of demand, such as retirement migrants, commuters and second home seekers.

But high rural house prices are not merely a reflection of strong demand from groups with the willingness and ability to pay more than lower-income groups. They also reflect several supply factors, such as a general absence of speculative building, high construction costs, and the tight planning restrictions on new development, reviewed in Chapter 3, and already proposed as a basis of class conflict in rural areas.

These supply factors are interrelated. One reason for the lack of speculative building has been the difficulty of obtaining sites large enough to interest the volume builders, due largely to planning restrictions which limit new building to infill sites within existing village boundaries. This has also prevented rural builders from enjoying the economies of scale associated with larger urban and suburban developments, so that unit construction costs have been higher. Construction costs are also often high because of planning conditions, requiring the use of expensive materials for example, as well as for reasons unrelated to planning, such as the costs of transporting materials, the more difficult topography, the shorter building season and the lack of local contractors and labour. These factors have led builders in many rural areas to cater for the most affluent, by building expensive luxury houses which justify the higher land costs and construction costs. Rogers (1976, 116) has observed that 'much private development is of a type, and consequently at a price, which effectively excludes many of those rural people who have the greatest need for housing'.

Newby's argument that this socially exclusive form of rural housing provision is a direct result of the post-war planning system and its mistaken assumptions about rural conservation has already been reviewed in Chapter 3, but it is worth emphasising.

> Since the Town and Country Planning Act of 1947 the granting of planning permission for rural housing has arguably been concerned with the visual quality of the countryside rather than with alleviating problems of housing need. By placing strict controls upon rural development these policies have brought about a planned scarcity of housing which, in the face of increasing demand, has made a rural house a desirable good with a premium price.... The population pressures on most of rural England are now those of increasing demand rather than those resulting from rural depopulation. The solution to this has not been to build even more houses to relieve the upward pressure on rents and prices but to impose even more stringent controls – conservation areas, village envelopes and so on. As prices inexorably rise, so the population which actually achieves its goal of a house in the country becomes more socially selective. Planning controls on rural housing have therefore become – in effect if not in intent – instruments of social exclusivity.
>
> (Newby 1980,186)

The dilemma for planners which Newby highlights, between maintaining an attractive landscape and providing housing for those in need, will be considered in detail in the case study of the Lake District in Chapter 6. At this stage it is sufficient to note that such restrictions on housing development result in costs and benefits which fall on different groups. Property owners gain from the general increase in land and house values consequent on any supply restriction, and they also gain from the preservation of their amenity. Urban dwellers share the benefit of a visually attractive countryside, although they may also suffer the consequences of higher urban densities and exported rural unemployment. Non-propertied groups in rural areas lose, however, in that their chances of buying houses in the countryside are diminished. To the extent that employers are discouraged, rural employment and wage rates will also be reduced, to their further disadvantage. In summary:

> The effect of such restrictive development policies has, then, not merely been to preserve the status quo in such areas but has been socially regressive: at one extreme those in expensively priced private housing have seen their environment retained or even enhanced; while at the other, low paid workers have been forced to leave the area.
>
> (Rose *et al.* 1979, 16)

Price, however, is not the only obstacle to house purchase in rural areas. Another major determinant of access to owner-occupation is the attitude of building societies to offering loan finance. The significance of the policies of building societies has been the subject of considerable research in urban areas (see especially Boddy 1980). Merrett (1982, 89) describes their lending policies as characterised by

> the systematic discrimination that follows from the role of the building societies as financial intermediaries, leading them to base their mortgage allocation criteria on the financial personality of the borrowers and not on the needs and existing housing situation of the household.

Discrimination has been found to exist in respect of applicants, dwellings and areas. Applicants tend to be favoured if they are young, salaried and in non-manual occupations. Older households receive shorter terms, with consequently higher repayments. Manual workers, with lower incomes, poor security of employment and more limited job prospects, are a bad risk (Short 1982, 128–9). Older properties, and properties in certain areas, are also unattractive to building societies. Their ideal property is a detached or semi-detached new house in a quiet residential suburb. Inner urban areas and remote rural areas may be thought quite unsuitable for lending, and may therefore be 'red-lined'. A survey of rural Scotland (Mackay and Laing 1982, 93) found a 'reluctance of the building societies to lend money for house purchase', attributed by local building society managers to policies set by their head offices.

The SDD (1979) study of rural Scotland examined this issue in some detail, and reached the following conclusions :

> Finding the finance is the main problem affecting the attempts of

> local people in rural areas to enter the owner-occupied sector. Building society lending in rural areas is fairly limited and tends to be very recent. Where there are signs of increasing activity (as in Orkney) it focusses mostly in newer housing in the higher price bracket and the effects on the local buyer (particularly the first time buyer) may be very small. Some old property is available and initially appears to be cheaper and thus more accessible to local people. However, in practice it is likely to be as inaccessible as much new housing since it is viewed less favourably by building societies and always attracts a lower proportion of loan assistance than new property.
>
> (SDD 1979, iv)

In those areas and on those properties not acceptable to building societies, the study found that local authority loans played a major part at that time, and it is unknown to what extent these funds have now been diverted to financing council house sales.

The decline of private renting

Many of the groups thus debarred from house purchase, either by low incomes or by ineligibility for loans, have traditionally relied upon the private rented sector in rural areas. This sector has tended to be larger in rural areas than in the towns, probably because of the widespread practice of farm, forestry and estate workers living in tied accommodation provided by their employers. Now, however, fewer and fewer houses are being made available to rent by private landlords, and the proportion of housing rented in this way has fallen from around 90 per cent in 1914 to only 11 per cent in Scotland and to 13 per cent in England and Wales in 1981. Table 5.7 indicates the scale of the decline, and the relative size of the private rented sector in rural areas.

Table 5.7 The decline of the private rented sector (percentages)

	1971	*1981*
England and Wales	22	13
Rural England[a]	24	10
Scotland	17	11
Rural Scotland	25	–

Sources: Censuses; Dunn *et al.* (1981, 39); DOE (1988b); SDD (1979, 60)
Note [a] The definition of rural England differs slightly between 1971 and 1981.

Gordon District Council (1978, 2) attributes this to 'a variety of reasons including the past lack of security of tenure in tied cottages, agricultural employment decline, and the provisions of the rent acts in determining the letting of such properties'. In some areas, this decline probably owes as much to the high prices which can be obtained from selling cottages to owner-occupiers, perhaps for commuting, holidays or retirement. Of course, these factors are interlinked. Gilder and McLaughlin (1978, 11) found in West Suffolk, for example, that the private tenancies which remained were mostly agricultural workers' tied cottages, and that the decline was caused mainly by the disposal of surplus rented agricultural housing to owner-occupiers.

In areas of particular landscape attractiveness, the higher prices and rents paid by holidaymakers for second homes and holiday cottages may encourage landlords to cease letting their houses to locals, as may the demand from commuters in areas of accessible countryside. The Peak District National Park has both these elements contributing to a decline of its private rented sector:

> At present the private rented sector forms the major source of housing for the local population and, with the high price of houses often ruling out the possibility of home ownership, this sector is playing an increasingly important role in local housing. 72% of those surveyed have tried to obtain rented accommodation. In view of the heavy reliance on this sector, the decline in the total private rented stock in the area from 2636 in 1961 to 1850 in 1971 represents a serious blow to the local community. The sale of previously rented properties once they fall vacant is an all too common trend which has unfortunate implications for local housing.
>
> (Penfold 1974, 15–16)

Similar trends were apparent in rural Scotland, demonstrating that it is not only in accessible, designated areas of attractive countryside that this tenure has declined. The SDD found that in respect of surplus farm cottages:

> Some older property in need of modernisation is sold off and in such cases much of it appears to go to outsiders. More often such property appears to be retained in the original ownership and is then let out, either as long-term second homes or as seasonal holiday cottages. There is evidence that in some rural areas surplus cottages on larger farms and estates are being left empty and gradually allowed to become derelict.
>
> (SDD 1979, 61)

The private rented stock which does remain in rural areas is very largely tied accommodation, chiefly for agricultural and distillery workers and other such groups. This has protected farm workers to a large extent from the effects of the housing shortages which confront other members of the rural population, albeit at the cost of some insecurity. Many of the tenants of agricultural tied cottages are approaching retirement (NACHA 1978, 6) and may have to vacate their cottages and seek council housing when they retire, reducing the private rented sector still further and placing even greater pressure on council housing.

The SDD's study concludes that,

> Although privately rented property accounts for a substantial part of the property in rural areas, access to it is limited....Those who are in this form of housing appear to be least able to get access to other housing either as owner-occupiers or in the public sector.
>
> (SDD 1979,v)

It continues:

> It appears that the balance between the different sectors in rural areas may be changing. Increasing costs of new property and either the high prices for improved property or the high cost of improving old property in poor condition is limiting accessibility in the owner-occupied sector. The private rented sector also appears to be declining or to be changing its pattern. As a result of both these factors the public sector may be looked to as the only alternative.
>
> (SDD 1979, 6)

The lack of council housing

Yet public provision of rented accommodation is very small in rural areas. Whereas in Scotland as a whole in 1971 more than half of all households lived in council housing, in rural Scotland only 38 per cent of households lived in this tenure. In England and Wales in 1981, 26 per cent of all households rented from the local authority, but in rural districts this proportion fell to only 17 per cent (DoE 1988b). Generally, the remoter the area the smaller the proportion of council houses provided. Mackay and Laing (1982) have estimated that in Scotland the proportion of dwellings which are council houses falls to 19 per cent in remote and to 11 per cent in very remote areas.

Evidence of the low rates of council building in rural areas exists for both England and Wales, and Scotland for the years 1968–73 (the only years for which these figures are published). Shucksmith (1981) presented a table showing permanent dwellings completed per thousand population in the rural districts compared with urban districts of England and Wales. Per capita council housebuilding in rural districts was only half the rate achieved in urban areas of England and Wales. At the same time, for every four and a half private houses built in rural districts (usually for owner-occupation) only one council house was built. This ratio compares unfavourably with urban areas, in which roughly equal numbers of private houses and council houses were completed.

Such low rates of council house provision were characteristic of the pre-reorganisation rural district authorities, and this has resulted in the very low stocks of public rented housing found in rural areas today. Even the figures quoted earlier, of 38 per cent in rural Scotland and 21 per cent in rural England in 1971, are probably overestimates of council house stocks in rural areas, since the 'rural districts' of 1971 included many small towns and suburban housing developments (reflected in the high figure for private completions too). Similarly, the 1981 figure of 17 per cent quoted for rural England and Wales refers to all settlements of less than 2,000 people. In the smaller villages of East Hampshire, for instance, the public sector has less than 10 per cent of the housing stock (Clark 1980). In South Oxfordshire, council housing represents 19 per cent of the total stock, but this is concentrated in the main towns while eighteen parishes have no council housing at all (Beazley *et al.* 1980, 21–3). A similar concentration of provision exists in Scotland. For example, 22 per cent of Orkney's households live in council housing, outside Kirkwall and Stromness the proportion falls to 7 per cent, and generally the proportion of housing provided by the local authority falls as one travels further from the main centres of population. This pattern is shown clearly by Martin's (1988) mapping of 1981 census figures for the proportion of council housing in each civil parish in Scotland.

Council house provision in rural areas is undoubtedly very low compared with urban council building. A person who joins the council waiting list in an urban area is far more likely to be offered a tenancy in the foreseeable future than someone waiting for rural council housing. It has already been noted that rural needs are generally found to be as great as those in urban areas, in that rural waiting lists are proportionately as long. Indeed, Shelter (1981) has shown that waiting lists in the Highlands are proportionately twice as long as those in Strathclyde. More broadly, Bramley (1989) has generated a needs index which suggests that the greatest shortfall in provision in England is in the non-metropolitan districts of the south.

Several of the reasons for this historically low provision of council houses in rural areas have already been mentioned in Chapters 3 and 4. Newby has argued that the rural local authorities before reorganisation lacked the political will to embark upon a rapid and expensive programme of housing construction, largely because the political control of rural councils lay with the farmers and landowners.

> Farmers and landowners continued to dominate rural areas, both as employers and as local councillors, and therefore as effective landlords of the housing stock within the range of farm workers: council housing, privately rented housing and tied cottages. Their desire to keep rates down (as major ratepayers) made them reluctant to build council houses. Farmworkers would instead be housed in tied cottages which had other advantages in addition to providing suitable accommodation, not least of which were the convenience of having workers on call, the greater stability tied housing conferred on the labour force, and the reinforcement of ties of dependency... Whether as ratepayers or as employers the farmers who ran the majority of rural councils found it more advantageous to provide tied housing for farm workers and to build the minimum number of local authority homes.
>
> (Newby 1980, 189–90)

Newby's findings in Suffolk are not confirmed in Norfolk, however, where a high proportion of council dwellings were built between the wars (Dickens *et al* 1985, 178–91). While the council was similarly dominated by farmers and landowners, there were other powerful political forces in the area:

> It was hard for the rural bourgeoisie to ignore the issues of poverty and poor housing, especially when faced with a discontented, sullen and hungry labour force. Moreover, this labour force was organised in trade union terms, and was beginning to vote Labour. Given this, the farmers were not too averse to the building of council houses....For the owners and managers it was far better that higher housing standards were paid for by the ratepayers in general, and by the taxpayer nationally, than for these to have been met by direct wage increases. The farmer may have lost some social control over the labourers if they moved out of tied accommodation, but the most valuable members of his workforce, such as the foreman and stockman, still tended to live close to the farm often remaining in tied cottages.
>
> (Dickens *et al.* 1985, 189–90)

In each case, the balance of political forces appears to have been crucial in determining the level of council provision relative to national trends.

Today, political factors continue to have importance, and the pressure on district councils is to build in urban areas, not rural. It has already been noted that Labour authorities in urban areas seek to direct building to brownfield sites as part of an urban development strategy and that Conservative shire authorities (under pressure from home owning residents) are equally keen to divert development pressures away from greenfield sites into the towns. The pressure on central government too, is to allocate resources to deprived inner-city areas rather than to the shires (ACC 1979; NACRT 1987). Within any district it is the housing problems of the towns which attract most attention, partly because of hidden need in the countryside, as mentioned, but also due to the concentration of voters. In addition, at both a local and at a national level it is the urban areas which contain the marginal constituencies, and where the electoral return on spending is likely to be highest.

Other reasons which were mentioned in Chapter 3 for the lack of council housing in rural areas included the land use planning policy of severely constraining the supply of land for housing development, the widely held belief in the social benefits of concentrating the population, and the economics of scale to be gained from building larger estates in the towns. Chapter 7 takes up the issue of central government's financial and administrative controls over rural council house building and it will become clear that central government's policies have played a major role in frustrating rural councils' attempts to build council houses, both through the capital allocations awarded and through the constraints placed on individual developments.

At this point, though, we return to the central theme of this chapter and attempt to pull together the strands pursued so far, so as to understand the overall pattern of advantage and disadvantage in the competition for access to housing in rural areas.

Access to owner-occupied housing has been found to be most difficult for people with low incomes and little wealth, especially if they are elderly or in manual occupations. Access to the declining private rented sector is largely restricted to farm and estate workers living in tied cottages, as fewer and fewer cottages are relet. Access to council housing is easiest for priority groups such as homeless persons and evicted farmworkers (in England): of the general needs groups, young married couples with children have the best chance of being rehoused, and the greatest difficulty appears to be experienced by elderly applicants and young single people. Overall, those aged over 60 and the unmarried under 30, without secure accommodation or a good income, are most disadvantaged in the search for housing because of their inability to buy, combined with the demise of the private rented sector and the low level of council house provision.

Access to housing improvement subsidies

In 1967, the Scottish Housing Advisory Committee reported that many families were 'inhabiting rural cottages in unbelievably squalid conditions, – without water, electricity or sanitation' (SHAC 1967, 23). They documented the unexpectedly large extent of the rural problem, and found that the landward areas of many counties appeared to have proportionately a bigger problem than the slums of inner Glasgow. They were highly critical of local councils, observing that 'the problem is not being tackled with the urgency which it demands' (SHAC 1967, 24).

The most recent surveys (SDD 1979; Mackay and Laing 1982) and the 1981 Census reveal that the rural areas of Scotland still contain the highest proportions of unfit housing, in contrast to England and Wales where 'rural housing problems undoubtedly remain...but they are more often those involving social equity and privilege than questions of poor sanitation and overcrowding' (Rogers 1976, 97). Such a dichotomy may be misleading, however, for it is one of the principal contentions of this chapter that the physical standard of rural housing is also a matter of social equity and privilege of access. Nevertheless, because poor rural housing is more prevalent in Scotland this section concentrates on Scottish conditions.

Inadequate housing is commonly defined in Scotland according to two alternative criteria. First, the census lists households which lack basic amenities: in 1971 these were hot water, a bath or shower, and a WC, but in 1981 the definition was narrowed to include only households lacking a bath and/or a WC. While this change of definition prevents a strict comparison between 1971 and 1981, it is believed to cause only a minor discrepancy. It can be seen from Table 5.8 that households lacking basic amenities were particularly prevalent in the remoter areas and islands. In Orkney, Shetland and the Western Isles the average proportion of households without these basic amenities in 1971 was between a quarter and a third, and it can be seen that the reduction in amenity deficiency has been least in these areas.

Table 5.8 Amenity deficiency (percentages)

	1971	*1981*	*% Change*
Scotland	13.5	2.8	–79
Rural and remote Scotland	12.0	3.1	–74
– rural	10.8	2.3	–78
– remote	14.8	4.7	–68
Urban Scotland	13.9	2.7	–81

(For definitions of rural and remote Scotland, see Appendix 1)

Further analysis by the SDD (1983a) indicates that most of these amenity deficient households are in the private sector, either rented from private landlords (47 per cent) or owner-occupied (37 per cent). In Scotland overall, 20 per cent of all private rented households lack basic amenities.

The census definition of amenity deficiency is narrow, however, and excludes many important characteristics of dwellings. The alternative criterion used in Scotland, dwellings 'below tolerable standard', is wider, taking account also of structural instability, dampness, lack of drainage, piped water, lighting and ventilation. Since this relates to the dwelling rather than to the household, some vacant or abandoned dwellings may be counted, particularly since survey information is typically ten or more years out of date. This also has the effect of failing to include any houses which have fallen below the tolerable standard since the last survey: where surveys have recently been completed, as in Badenoch and Strathspey, Ross and Cromarty and Nithsdale, the numbers of BTS houses have grown markedly, not decreased (Shelter/Rural Forum 1988). Nevertheless, the available statistics (such as they are) confirm the general pattern of a high rural incidence, and of a slower rate of improvement in rural areas, as shown in Table 5.9.

The reasons for the slower rate of improvement in remote areas are described in Shucksmith (1984, 5–8). In summary, the principal means by which BTS housing in urban areas has been improved, the area-based Housing Action Area (HAA) designation, has proved inappropriate and cumbersome in a rural context. Thus, during 1978–82, only 1,060 rural dwellings were improved by this means (2.5 per cent of BTS stock) compared to over 10,000 dwellings in urban Scotland (9 per cent of BTS stock). Neither have the housing associations, the principal instrument for implementing improvements in urban HAAs, been effective at rural improvement: outside Argyll and Bute (an honourable exception), only 189 units had been rehabilitated by housing associations in rural and remote Scotland up to 1988, compared to over 18,000 in the four cities, although 46 per cent of BTS dwellings are in these former areas. Further, the government's approved expense limit for improvement and repair costs takes no account of the higher costs incurred in rural, and especially in remote, areas. The SDD conducted a survey of BTS housing in five rural districts in 1977 and found that the costs of repairs (1982 prices) varied from £7,000 in Ettrick and Lauderdale and £9,000 in Moray to over £21,000 in the traditional crofting areas (SDD 1983a).

Perhaps most relevant to the theme of differential access, however, is knowledge of which households are most likely to be living in below tolerable accommodation. Two features are particularly noteworthy. First, the tendency of BTS housing to be associated with crofting tenure suggests that crofters as a group may be relatively disadvantaged.

Second, the tendency for BTS houses to be occupied by elderly people suggests that the elderly may also face particular disadvantage.

Table 5.9 Dwellings below tolerable standard (percentages)

	1976	*1984*	*% Change*
Scotland	8.0	3.5	–56
Rural and remote Scotland	9.0	5.5	–39
– remote	13.0	8.7	–33
– Western Isles	31.0	21.0	–32
– Skye and Lochalsh	24.0	13.0	–46
– Argyll and Bute	21.0	15.0	–29
– Orkney	29.0	15.0	–48
– Shetland	19.0	12.0	–37
– Ross and Cromarty	12.0	14.0	+17

The elderly

Considering the position of elderly people in rural areas, the SDD (1979) found that in rural Scotland 'The occupants of sub-standard housing tend to be small, elderly households with low incomes; 75% of BTS households comprise one or two persons; 63% of BTS dwellings are occupied by households headed by a person aged over 60.' The problem of BTS housing in rural areas is therefore not merely a physical problem of unfit housing: it is a reflection of the inability, or perhaps the unwillingness, of elderly people to maintain and repair their houses or to move to better accommodation. As we have already noted in this chapter, elderly people with low incomes are less likely to be able to obtain loans from building societies, or other agencies, with which to improve their houses or with which to finance the purchase of another house. Equally, they may well be highly resistant to any suggestion that their houses be improved, with all the disruption that entails. They may therefore be trapped in sub-standard dwellings with no real prospect of improvement or of moving elsewhere.

Crofters

Crofting tenure, while bringing many advantages to the crofting community, also tends to disadvantage certain categories of crofters. Because of the special nature of crofting tenure, by which a crofter rents his land from the landlord and builds his own house upon it, crofters normally have no heritable security to offer to a bank or building society as collateral for a loan. Most importantly, crofters are ineligible for loan assistance towards house improvement from the local housing authority. Instead, a special scheme of grants and loans is operated by the Department of Agriculture and Fisheries for Scotland (DAFS) for the erection, improvement and rebuilding of crofter houses. For improvements, a grant of £1,500 and a loan of up to £8,000 at 7 per cent over twenty years are available. Generally, DAFS will be the only source of loans available to crofters. The scheme is described in detail in Shucksmith (1987a).

There are major difficulties with this scheme, of which the most notable are that the subsidies for home improvement thus available compare unfavourably with those available to non-crofters from local authorities, and have failed to keep up with the costs of improvement and repair work. Additionally the subsidies tend to favour those who have regular employment, or whose crofts are large enough to provide a comfortable income, rather than the unemployed, the elderly and crofters receiving only low incomes from their on-farm and off-farm activities. This is also examined further in Shucksmith (1990, forthcoming), which suggests that the major determinant of poor housing conditions in crofting areas is the inability of poorer crofters to make use of these subsidies. In this survey of households in Uist, a very high proportion of crofters living in BTS housing were found to have no source of income other than their pension, social security payments or the croft itself. A strong relationship was apparent between income, employment and the implementation of improvement work. Households which had sufficient means to improve their housing conditions tended to do so.

BTS housing and housing without basic amenities are therefore symptoms of the difficulties faced by crofters without regular incomes and by elderly people in gaining access to the subsidies which other groups have been able to use to improve, repair and maintain their houses. In each of these cases, the individuals concerned generally have no option but to continue to live in sub-standard housing. Housing in poor physical condition thus reflects real disadvantages faced by particular groups which have poor access to housing improvement subsidies.

Conclusion

The political economy framework developed in Chapters 2 to 4 has suggested an empirical focus on class conflict, conflicting policy objectives and competition in housing consumption. In particular, while the analysis of conflict was seen to centre on domestic property classes, the investigation of inequities in housing consumption implied a central concern with the means of access to housing. This chapter has considered empirical evidence of differential access to housing in rural Britain, and this has allowed a general pattern of disadvantage to be identified, notwithstanding the variation in circumstances between different areas. It is clear, even from the partial evidence available, that the elderly, the young and others in insecure rented accommodation are disadvantaged in terms of housing consumption, relative to more prosperous groups within rural Britain, in that they find it increasingly difficult either to find accommodation or to obtain the improvement grants necessary to bring BTS housing up to standard. In many cases these disadvantages will extend to a lack of opportunity for domestic capital accumulation, where housing consumption classes correlate with domestic property classes.

This chapter has not pursued the issues of conflicting class interests and conflicting policy objectives over land release for new building, and yet these are clearly interrelated with distributional outcomes. The case study which follows, of the Lake District, allows these distributional issues to be explored in greater depth. Most pertinently it will also be possible to examine the interrelationship between the two principal themes of this book, the conflicting objectives of rural housing policy and the associated distributional outcomes for different social groups.

Appendix to chapter 5: definition of rural and remote Scotland

The problem of defining rurality, and of rural heterogeneity, has been recognised and accorded lengthy discussion (cf. Pacione 1984 and Phillips and Williams 1983). Some definitions have been based on land use and population density criteria, although these have been criticised by Whitby and Willis (1978) for failing to consider the complex social structure of rural areas. There are also problems of the level of analysis: a classification of EDs would reveal localised housing circumstances, but the level of policy (and most published information) is the district.

The most sophisticated approach has been that of Cloke (1977, 1979) who derived an index of rurality from multivariate analysis of sixteen variables measuring remoteness, population, housing, occupation and migration characteristics of pre-1974 districts. This work only extends to England and Wales, unfortunately. Webber and Craig's (1978) classification, based on a cluster analysis of 1971 census data, has been used to summarise 1981 data; but although this extends to Scotland it fails to reflect the diversity of housing circumstances within rural Scotland (apart from abberations such as the inclusion of Aberdeen City as rural).

More helpful in this respect, although still based on 1971 data, is the Scottish Office's Rural Indicators Study (Burbridge and Robertson

1978), which examined district and ED level data and so suggested a classification of districts. While not ideal, this allows some disaggregation of rural areas within Scotland, and may be justified on pragmatic grounds. The crucial variables were not housing related, but 1971 population density and employment in agriculture, forestry and fishing. The use of districts as the unit of classification at least accords with the local housing authorities (the units of policy) and thus with published sources on housing. The classification has subsequently been used by the SDD to distinguish between 'dispersed rural settlements' and 'rural areas with small towns' (SDD 1979, 1983a, 1983b, 1986). These two definitions are adopted by this author, and for brevity are referred to as 'remote' and 'rural' respectively, though it should be realised that there may be considerable diversity even within a single district: for example, Gordon District, while it is classified as 'remote', contains both remote glens and dormitory suburbs for Aberdeen. The distribution of rural and remote districts, thus defined, is shown in Figure 5.1.

The classification is adopted as a pragmatic means of organising and presenting pre-existing published information. Its shortcomings reflect the lack of published housing data at sub-district level. It is inevitably crude, and to some extent misleading: however, housing market information is neither collected nor published at the more localised level which would be desirable.

Figure 5.1: Rural and remote Scotland.

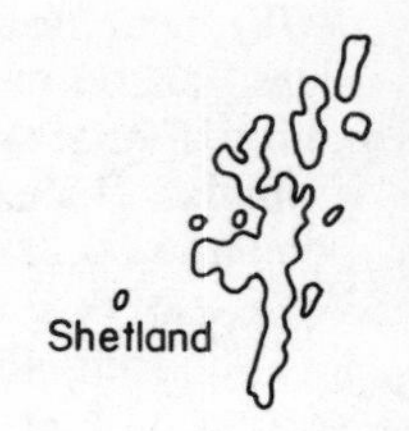

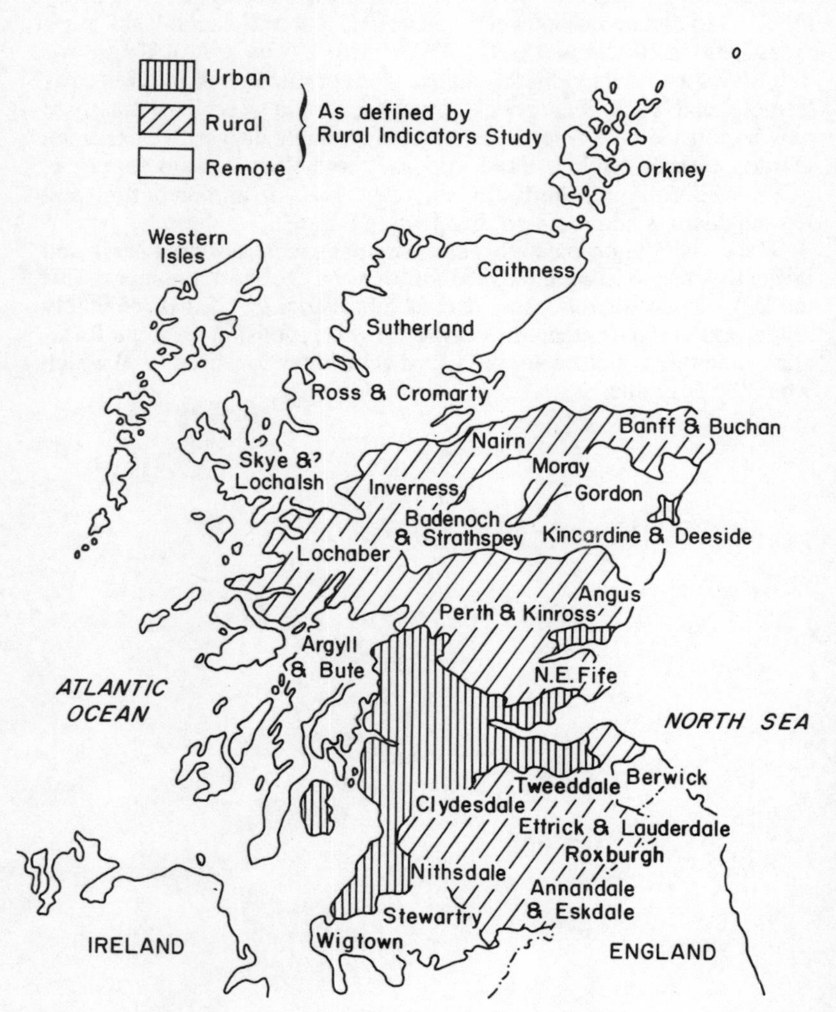

Chapter six

The Lake District – a case study

Introduction

One of the twin themes of this book is the conflict of public policy objectives between landscape protection and the release of land for housing. In an area of great landscape attractiveness, such as the Lake District, this conflict is particularly acute as attempts are made to protect the landscape while remaining responsive to the housing needs of local people. For this reason, the Lake District offers a highly appropriate case study, allowing the effects of this conflict of objectives to be analysed in depth. Further, the effects of giving priority to landscape preservation over housing objectives are likely to be distributionally uneven, with property owners generally benefiting while poorer groups find their housing disadvantage reinforced, as suggested in Chapter 3. This case study therefore affords the opportunity of also pursuing the equity theme, developed in Chapters 4 and 5, and of examining the ways in which equity and efficiency interrelate in practice.

The principal reason for choosing the Lake District as a case study, however, lies in the opportunity it affords to re-examine the innovatory and still controversial policy response of the local planning authority, which attempted to resolve this conflict of public policy objectives and at the same time to influence distributional outcomes in favour of one group (young locals) who faced housing disadvantage. This policy response has already been analysed in detail (Shucksmith 1981; G. Clark 1982), but there are several reasons for re-examining it. First, the empirical basis of the earlier work was somewhat inconclusive, since too little time had elapsed since the introduction of the policy to assess its effects on observed house prices: more data are now available. Second, some criticisms of the earlier work have appeared (Loughlin 1984; Capstick 1987) and these are discussed and rebutted. Third, the current government policy initiative to encourage building in villages by housing associations appears to use similar instruments to those criticised as counter-productive and inequitable in the Lake District experiment (Shucksmith 1981), and it is therefore very relevant at the present time to examine the degree to which such criticisms might also be applied to the current NAC Rural Trust initiative.

Land use conflicts have a long history in the English Lake District. Since the romantic movement of the eighteenth century, the area has been generally regarded as picturesque and attractive, and the poet, William Wordsworth, in 1810 suggested that the district might be regarded as 'a sort of national property, in which every man has a right and interest who has an eye to perceive and a heart to enjoy' (1810, 92). The concept of a public good clearly has long antecedents.

In the nineteenth century, the idea of landscape preservation came into conflict with many other land uses. Most notable perhaps was the arrival of the railway and the consequent construction of large villas along the shores of Windermere and other lakes. Further conflicts arose over the building of a water supply reservoir at Thirlmere in 1890 and over various proposals for mineral extraction. A number of pressure groups and other bodies were founded during this period, including (most notably) the National Trust. By 1902 the National Trust had begun to acquire land 'for the nation' in the central Lake District.

But conflicts were not only between development proposals and preservation. The pressure of visitors began to cause conflict as walkers crossed farmers' enclosed land to gain access to the open fells. Visitors and local business interests also came into conflict with farmers and preservationists as efforts were made 'to publicise and open out the Lake District, with all that it involved' (Berry and Beard 1980, 3).

Today, the major competing interests for the use of land in the Lake District are recognised in statute. The area is designated as a national park. A national park authority, the Lake District Special Planning Board (LDSPB), exists to pursue three fundamental statutory objectives, which reflect the competing claims on land in the national park:

1. to preserve and enhance the natural beauty of the landscape;
2. to promote public enjoyment of the park; and
3. in pursuit of these two principal objectives, to have due regard to the economic and social interests of the local communities.

As argued in Chapter 3, the pursuit of each of these objectives may be viewed from the point of view of the economist as justifiable public intervention in the land market, on the ground that each promotes a land use which would be undervalued in the free market. It was suggested that the economic and social welfare of local communities in an area like the Lake District is a form of producer good, essential to the production of two public goods, landscape preservation and public enjoyment. Land use for the economic and social benefit of local communities may thus be encouraged not only from an equity perspective but also on efficiency grounds. This argument highlights the conflicting objectives of policy relating to rural housing, and it is the purpose of this case study to explore these in greater detail. Some background knowledge is necessary, and readers who are unfamiliar with the Lake District housing market and the structure of local government in the national park area may find a brief account in Shucksmith (1981, 1987).

It is worth noting at this juncture, however, the composition of the LDSPB. The board consists of thirty members: sixteen are appointed by Cumbria County Council, four by the district councils, and ten by the Secretary of State for the Environment. These ministerial nominees are appointed to represent the national interest, and this has tended to mean a conservation or outdoor recreation interest. Capstick (1980, 63) conducted a survey of ministerial appointees on the pre-1974 board, and 'all were clear that their past activities in conservation or outdoor recreation movements had led to their nomination, and all had developed strongly protective attitudes towards the landscape'. Since the appointed members of the LDSPB typically served for much longer periods than county nominees, and held most of the offices, 'opinion is unanimous that the most influential members were those nominated by the Minister' (1980, 68). This would seem to apply equally to the post-1974 LDSPB, the appointees to which have often been landowners, farmers or members of amenity groups. Thus, the board has been dominated by the conservation and recreation interest groups since its inception.

The influence over the LDSPB of such interest groups, and of the Friends of the Lake District in particular, is apparent from the response of the LDSPB's then chairman to a parliamentary committee:

> The chief agent of the National Trust is on my board, there are various people to do with the Ramblers. All our 9 appointed members have some considerable interest in the main interests of the Park....We get on very well with the Secretary of the Friends of the Lake District, he is not on the board now but he was in the old days....The main interests are represented, the NFU and the CLA..., we also have Dr Halliday for nature conservation. I think we are well represented.
>
> (HMSO 1976, 378)

Brotherton (1981) has argued that the success of the board in conserving the Lake District landscape is largely a consequence of the support it has enjoyed from these powerful and articulate park pressure groups. The reliance of the board on these groups for its legitimation must be borne in mind when considering its policies.

The incidence of housing disadvantage

The concept of housing disadvantage has already been discussed at length. This section assembles and analyses the evidence available of the incidence of housing disadvantage in the Lake District, again based to some extent on waiting list information supplemented by a discussion of hidden needs, but also through an examination of public and private sector allocation processes.

In the Lake District the most significant dimension of the housing market is that of tenure. The mechanisms for access to owner-occupation are quite different from those for access to rented accommodation, whether public or private sector, as argued in Chapter 3. In the Lake District over half the permanent households are owner-occupiers, and this proportion increased from 52 per cent to 58 per cent during 1971–81, in line with the national trend. The private rented sector declined sharply from 34 per cent to 25 per cent, while the small public sector remained at a fairly constant level (17 per cent), prior to the effects of council house sales post-1980 which led to a decline in the councils' stock from 2,264 houses to 1,826 in 1986 (a net loss of 20 per cent).

This account of the tenurial structure of the Lake District is complicated by the presence of holiday cottages and second homes. The 1981 Census revealed that 16.4 per cent of dwellings in the Lake District were second homes or holiday cottages, with the proportion rising to over 35 per cent in particularly popular parishes like Langdale and Patterdale. If second homes, holiday cottages and vacant dwellings are distinguished from the effective stock, then the tenure distribution is as shown in Table 6.1.

This highlights the small proportion of the housing stock available to the permanent population to rent, and points towards the difficulties which lower income groups may therefore experience.

Table 6.1 Housing tenure in the Lake District, 1981, distinguishing non-effective stock (percentages)

Owner-occupied	43.9
Council houses	12.6
Private rented	18.8
Second/holiday homes	16.4
Absent/vacant	8.2

Competition for housing

Concern about the housing market in the Lake District is often couched in terms of young people being forced to leave the area, and an ageing of the population. Overall, the park's total population has changed very little in the last thirty years. However, this stable aggregate conceals substantial changes in the composition of the park's population, with a declining number of children and an increasing number of retired people. In 1981, for example, the ratio of retired people to population of working age was 41:100, relative to a national average of 29:100. The number of persons aged over 60 increased by 10.5 per cent between 1971 and 1981 largely due to retirement migration to the larger and more accessible settlements (Capstick 1987). In particular, the accessible south-eastern area between Windermere and the motorway has experienced a large increase in its retired population, as have the towns of Windermere and Bowness to a lesser extent. The greatest loss of young people has been in the Eden District and Western Dales areas (LDSPB internal working paper) each of which is experiencing an aggregate population loss, and has a more restricted range of employment opportunities and little public sector housing.

Second home buyers have generally been blamed for displacing young people in the Lake District and elsewhere. Shucksmith (1981) suggested that a more potent source of external demand for housing in the Lake District was the retirement migrant, often with a greater ability to pay following the sale of his previous house. These findings are borne out, not only by the demographic trends, but by a more recent survey of estate agents, in which 'experienced agents in all parts of the Park were interviewed' (Capstick 1987, 63). This found that 'the second home market still exists but is less important than it was' and that 'demand from outside the Park is mainly from people moving into the area for retirement'. 'Total outside demand varies, but in popular and accessible parishes,... it may reach 50% to 60%.' (p. 64). This estimate was confirmed by Capstick's analysis of the origin of solicitors' searches in the land registries in 1985, which also indicated that the areas most sought after by outside buyers were the central Lake District, the Keswick area and the valleys around Ullswater. Outsiders were less interested in the northern and eastern peripheries. The survey also confirmed that those local people who were able to buy houses in the area were not first-time buyers but middle-aged people with high incomes.

In Shucksmith's (1981) survey, estate agents were asked whether locals and newcomers often compete for similar property: in retrospect it would have been more pertinent to have asked to what extent local first-time buyers compete with newcomers for the same property. Further evidence on this point comes from a survey of the occupiers of several recently built housing estates in the Lake District (LDSPB 1980b), which shows quite conclusively the existence of direct competition between locals and holiday and retirement home buyers. The results of this survey are summarised in Table 6.2. Certain caveats apply, however, since the LDSPB admitted at the public examination of the structure plan in 1980 that this sample was drawn selectively to illustrate the extent of holiday and retirement ownership: it is not a random sample.

Table 6.2 Occupiers of selected housing estates (percentages)

	Ten recent housing developments	*Eleven pre-1900 stone-built terraces*
Holiday cottage/second home	26	23
Retirement home	40	43
Local occupier	34	34
Total houses in sample	344	100

Source: calculated from data in LDSPB (1980b)

Nevertheless, it demonstrates that competition for housing exists between local buyers and holiday and retirement home buyers at both ends of the housing market within major settlements. The recent developments surveyed ranged from a small scheme of ten bungalows and another of twelve flats to a large estate of eighty nine detached and semi-detached houses. The traditional terraces represent the cheapest housing available for purchase in the national park and yet their pattern of occupancy was little different from that of the newer developments. The LDSPB concluded that there was no type of private housing development which would meet local demand without competition from holiday and retirement home buyers (LDSPB 1980b, Annex 1).

Information from waiting lists

The most readily available, if imperfect, source of information about housing disadvantage in the Lake District is the council house waiting list. The information from waiting lists gives only a partial picture of housing disadvantage, of course, and the existence of hidden needs in the Lake District is apparent. However, the waiting list applicants' characteristics reveal some of the dimensions of housing disadvantage in the area (see Shucksmith 1981; G. Clark 1982; Capstick 1987). The main groups on waiting lists were families in insecure or shared accommodation and elderly people living in unsuitable houses. According to Capstick (1987, 35), the length of waiting lists has changed little since 1980, but within them 'the emphasis has moved from families to old people'.

Capstick (1987) found that council building in the park is now so little that it would take six to ten years merely to house the current priority applicants. She noted that between 1983 and 1986 there was no relet of any kind in Ambleside and that in Hawkshead there has been no relet of a family house since 1977, despite long waiting lists. Needs were unlikely to be expressed in such circumstances. Bennett (1976, 39) identified several groups who were unlikely to join a waiting list while nevertheless exhibiting hidden need, including 'those who cannot wait because they need accommodation when they first get married, people without children, single people, and people who have no council houses in their village'. Eligibility restrictions for entry to the waiting list may add further groups to this list, further detracting from the utility of the waiting list as an indicator of housing disadvantage.

Despite these shortcomings, some idea of the incidence of housing disadvantage can be built up from analysing the waiting list in an attempt to observe need directly, and attempting to counter its deficiencies by accumulating anecdotal evidence. An alternative approach, however, might focus on the mechanisms of housing allocation, and such an approach is now attempted. A further approach, involving interview surveys, would attempt to follow the housing careers of a sample of households, but this is not pursued here.

Mechanisms of housing allocation

It is already apparent that low–income and even middle–income groups who do not already own property are denied entry to the market for owner-occupied housing in the Lake District because of the area's attractiveness, which justifies tight planning controls on the supply of new housing at the same time as stimulating more prosperous groups to seek retirement and holiday homes in the area. Together these result in high house prices. Shucksmith (1981) compared house prices in 1979 with the then distribution of income in Cumbria, concluding that perhaps 70 per cent of the population were too poor to buy any form of accommodation in the Lake District. Those who had an income sufficient to buy were probably already owner-occupiers. By 1985, the LDSPB (1985, 199) reported that the cheapest terraced cottage would typically cost at least £28,000. Capstick (1987, 29) considered Lake

District income data from local job centres and building society offices, as well as national data from the New Earnings Survey, and she confirms that

> many working people in the Lake District and its periphery have low incomes and can not hope to be in the open housing market even in the larger and more urban settlements of the National Park.... What we learn from this is that the rented sector will be needed in the Lake District since many working people will not be able to enter the open market.

This is an important point, because so much of policy and of the literature have concentrated on the high price of house purchase, when the shortage of housing to rent is arguably more critical for lower income groups.

South Lakeland District Council (1977, 2) considered the private rented sector in its area and reported that 'relets in country areas are so rare that a new tenancy in a particular locality may literally not be possible for decades'. The only people likely to find private rented housing are farm and estate workers, such as those employed by the National Trust, and those workers in tourism and catering establishments who are offered tied accommodation. Therefore, despite the size of the private rented sector, its contribution to future housing requirements is likely to be small.

The vast majority of lower-income households are thus reliant upon gaining access to the small stock of council housing if they are to find homes within the Lake District. Yet it is not possible to say with confidence which groups find difficulty in this because no full comparison of applications with allocation outcomes has been undertaken. Unfortunately, neither G. Clark's (1982) nor Capstick's (1987) analysis of waiting lists in the Lake District posed such questions and they provide no information comparable to that reviewed in Chapter 5. G. Clark (1982, 81) at least notes that:

> The low proportion of people who believed themselves in need of a council house and who were rehoused by the council is rather perturbing. Rapid rehousing by the council may only be possible because two-thirds of applicants find their own houses, die or migrate from the Lake District. The question of which of those who cannot buy a house shall be allocated a council house seems relevant in this context. The relationship between the number of council houses, their speed of allocation to applicants and depopulation of the low incomes group from the countryside clearly needs some investigation.

His analysis of allocation outcomes in the Allerdale part of the national park reveals only (1982, 81) that 'housing points are not a good predictor of the time needed to obtain a council house', and that Allerdale District Council takes several other factors into account, including 'less easily quantifiable information on the person's exact situation, the urgency for rehousing and the benefits to the local authority from rehousing particular individuals'. This system he endorses uncritically as 'more flexible, less predictable and more humane, relying on broader criteria of merit' without refering to the well-known potential for patronage and abuse inherent in such a system. Nor does he identify which groups are allocated council houses by this system, except to note that pensioners and families appear more successful than single people, and that tenants fare better than owner-occupiers or lodgers. Capstick (1987, 32) examined the points system of each authority in the national park and found that:

> The rank order of need was the same in all districts. This was:
>
> 1) homeless families or those under immediate threat of homelessness;
> 2) families with children who share living accommodation;
> 3) principal wage earners of households who travel 15 miles or more to work;
> 4) families separated because of housing difficulties.
>
> Housing for the elderly comes into a separate category.

She suggests that there is an increasing need for elderly persons' housing, but that this is being matched by councils' preference 'to build special housing for old people because this is exempt from the tenants' right to buy' (p. 50). She identifies childless couples and single people as the groups with

> no priority for council housing; these were found living with their families, separately, or in caravans, or in expensive... accommodation at long distances from their work... The one case found of a single person living in a tent is the extreme expression of a deprivation which, though at any one time existing in small numbers in each parish, is real and continuing.
>
> (Capstick 1987, 99–100)

More detailed information can be gained from Allerdale District Council (1978), which sets out the council's allocation scheme and discusses categories of need. The council gives special priority to families who have to live with relatives (living-in), or in overcrowded conditions or in bad housing. In contrast, the needs of single people are regarded as 'not as pressing' as those of families, even when in 'extremely difficult circumstances', and single people are accordingly given a low priority in the allocation system. The increasing need for smaller units for elderly people is recognised, although meeting this need is seen to depend on new building given the shortage of smaller houses. These priorities are confirmed by a comparison of waiting lists and allocation outcomes in Allerdale's area of the national park, as shown in Table 6.3. It can be seen that families are favoured, and that the elderly and especially the single are disadvantaged by the allocation process.

Table 6.3 Waiting list and allocations in northwest Lake District, 1979 (percentages)

		Waiting list	*Allocations*
Families:	Living in or caravan	9	35
	Tenant/owner-occupier	25	32
Single:	Living in or caravan	15	1
	Tenant/owner-occupier	7	2
OAP	Living in or caravan	7	6
	Tenant/owner-occupier	37	23

Source: J. Blackie (TRRU), personal communication.

An approximate picture of housing disadvantage in the Lake District has thus been established. It can be concluded that the more prosperous sections of society have the advantage of access to owner-occupation, and that private rented housing is now only accessible to workers in occupations which offer tied accommodation. Of the poorer groups which remain, the pattern of housing disadvantage depends to a large extent on the processes of public sector housing provision and allocation. In the Allerdale area of the national park, these processes favour families who seek housing in the main town, Keswick, while discriminating against young single people in particular, and against those who wish to remain in their village. The shortage of smaller council houses tends to disadvantage elderly applicants somewhat. While Capstick (1987) seems to confirm this pattern of disadvantage in other areas of the park, her analysis is imprecise and further analysis is required before a full account of housing disadvantage can be given.

The policies of local housing authorities

Since the district councils are the local housing authorities for the Lake District area, with responsibility for meeting housing needs, they must be held prima facie responsible for the housing difficulties faced by certain groups. Bennett (1977, 28) notes the low provision of council houses in the national park and indicts the local authorities for their lack of imagination and political will. The Cumbria Countryside Conference (1979) explained the lack of council building in terms of higher rural building costs and a failure to observe and react to hidden needs.

However, the explanation may have as much to do with the historical legacy of a neglect of council housebuilding in the inter-war period and with financial constraints imposed upon local housing authorities by central government since reorganisation in 1974. The proportion of council houses in the national park in 1951, after all, was only 2 per cent (G. Clark 1982, 63), compared to a national percentage of 18 per cent. Between 1951 and 1976, despite all the factors militating against rural council house provision, 40 per cent of all new houses built in the national park were in the public sector (LDSPB 1978b, 151), although Shucksmith (1981) has shown that during the last seven years of the old authorities (1967–73) only 22 per cent of new houses were council houses, a similar proportion to that pertaining since 1976. This suggests that the low level of council housing in the national park today derives principally from failings prior to 1951 and after 1969.

Capstick (1987, 34) observes that each council, upon reorganisation

> inherited a stock of housing whose state and location owed nothing to the new district's own assessment of priorities. The predecessor authorities varied in their willingness to spend on housing and in their location policies. North Lonsdale Rural District, for example, left a reasonable scatter of small groups of council houses in the villages in the southern part of the park, as well as a heavy concentration in Coniston; the former Wigton Rural District did not provide housing in even such a sizeable village as Caldbeck. The small settlements are, still, worse served in Allerdale than in the other districts.

More recent failings can be gauged from considering the period 1977 to 1984 during which eighty seven council houses and ninety five housing association houses were built in the national park out of a total of around 1,000 new houses: together these amount to only about 20 per cent of new houses. The extent to which the post-1974 housing authorities are themselves culpable for this dearth of council housebuilding, and the extent to which they have been prevented from building council houses by the financial restrictions progressively imposed by central government since 1975 is considered in detail in the next chapter. Capstick (1987, 54) shows that 11.5 per cent of the council houses in the Lake District were sold to their tenants under the 'right to buy' between 1981–5, but that the high prices (even after discount) tended to discourage sales in the most attractive areas. Nevertheless, sales have exceeded completions, with a net loss of stock of 20 per cent between 1980 and 1986. The main constraint facing local authorities has been the limit, imposed by central government, on borrowing to finance construction and the extent to which this 'capital allocation' has been cut back since 1981 for the four districts concerned is shown in Table 6.4. In sum, capital allocations have been cut by 54 per cent during 1981–7. Capstick (1987, 119) is confident that 'had finance been available, [the housing authorities] could well by the present time have reduced somewhat the length of the waiting period for rented housing in the Lake District'.

Table 6.4 Capital allocations to Lake District housing authorities (at 1985–6 prices)

1981–2	£11.80m
1982–3	£10.93m
1987–8	£5.37m

The shortage of local authority housing is certainly a central element in the development of current housing problems facing disadvantaged groups in the Lake District. Whereas elsewhere, prior to the 'right to buy', council housebuilding had largely offset the decline of the private rented sector, in the Lake District the poorer households have seen their opportunities diminished by the disappearance of rented housing. The private rented sector in this area was particularly important in fulfilling this function, but this is in rapid decline: instead of expanding to fill this gap in provision, council housing in the Lake District remains significantly lower than in other rural areas and is now declining in absolute numbers. However, as suggested, this shortage is due only partly to current failings on the part of the post-1974 housing authorities, and both central government policies and low historical levels of provision must bear part of the blame.

In terms of policy, while the local housing authorities remained either hamstrung by central government constraints or preoccupied by housing needs outside the park, depending on one's point of view, the planning authority was confronted by the central dilemma identified in Chapter 3. Should planning controls be tightened in the interests of landscape protection as the limits of acceptable development are approached; or should planning controls be loosened in the interests of relieving the pressure on the housing market and thus, perhaps, helping disadvantaged groups find accommodation inside the park? It is to the LDSPB's policy response to this apparently irresolvable dilemma that the discussion now turns.

The Lake District Special Planning Board's policies

Since the board's inception, in 1974, it has been extremely concerned about housing for local people within the national park. The motive force behind this concern appears to have been the board's realisation 'that the difficulties which young people found in obtaining housing were the chief concern of the inhabitants of the Lake District' (Capstick 1987, 133). While the board's primary aim is to preserve and enhance the landscape, it is arguable that the board derives its authority and legitimation not only from the support of powerful and articulate preservationist pressure groups, such as the Friends of the Lake District (Brotherton 1981), but also from the more tacit support of the people living in the national park. For this reason,

> it was very difficult to sit back and complain that nothing could be done, when the need for houses for local people was raised at every public meeting which the Board held.
>
> (LDSPB 1978a, 21)

Therefore, in 1977 the board announced a new policy which 'will restrict completely all new development to that which can be shown to satisfy a local need' (LDSPB 1977b).

> The villages and towns of the National Park could not be permitted to expand for ever unless the farming industry and the landscape were to suffer irreparably and in many areas the amount of land suitable for housing was beginning to dry up.
>
> (LDSPB 1977d)

The board therefore asked any applicant for a residential planning consent to sign an agreement under section 52 of the Town and Country Planning Act 1971, limiting the future occupancy of the housing to people employed or about to be employed locally, or retired from local employment. The origins and evolution of the 'homes for locals' policy, or the 'section 52 policy', as it became known, are described in Shucksmith (1981). The exact wording of the policy contained in the Cumbria and Lake District joint structure plan submitted in 1980 was as follows:

> The Board will seek means by which all further housing development can be retained for occupation by local people as full-time residents, except in the case of divisions of existing dwellings or where exceptional site factors apply.
>
> (LDSPB 1980a, 55)

The board maintained that the policy would

> ensure that the diminishing number of sites where new housing will be acceptable are used to maintain rural life.
>
> (LDSPB 1978a, 22)

This policy was identified by the Secretary of State as a matter for debate at the examination in public of the structure plan in September 1980, where the policy was closely scrutinised. The examination panel's report and the Secretary of State's proposed modifications were published in August 1981. The panel concluded that:

> There is in the Lake District National Park an acknowledged problem of attempting to meet the housing needs of local people in a situation where there is an extraordinarily great demand for that housing from people from elsewhere in the country. We can see no adequate solution likely in any of the alternative courses of action which have been canvassed before us although a small contribution can be made through the use of Housing Act powers. Because the LDSPB's policies have only been in operation for a very short time, it is difficult to reach any firm conclusion on the likely effects for good or for ill but because the problem is so extreme and so intractable, it is our conclusion that the Board's policies should at least be tried.
>
> (DoE 1981b, 17)

Despite this qualified support, the Secretary of State proposed to delete the policy, because in his view,

> it would be an unreasonable use of planning powers to attempt to ensure that houses should only be occupied by persons who are already living in the locality. The objectives of the authority are best served by the use of their powers under the Housing Acts...[sic]

Further,

> Planning is concerned with the manner of the use of land, not the identity or merits of the occupiers. Planning permission for a particular use of land otherwise suitable for that use cannot appropriately be refused simply because the planning authority wish to restrict the user.
>
> (DoE 1981a, 10–11)

The LDSPB (1982, 10) was encouraged by the panel's support and hoped that the Secretary of State's opposition was based on a misunderstanding; the board undertook vigorous lobbying, enlisting the support of the then Chief Whip, Michael Jopling, and the Home

Secretary, William Whitelaw, whose constituencies included parts of the national park (*Planning*, 3 December 1982). In the event, this political pressure merely delayed the decision.

In December 1983, the Secretary of State confirmed his decision to delete the board's policy from the structure plan, arguing that

> it is not in general desirable to seek through planning restrictions to control the disposal of private houses. Planning control is essentially concerned with the use and development of land; not the identity and/or characteristics of the user; and only in very exceptional circumstances is it legitimate, in his view, to deprive a householder of his normal rights to sell or let his house to whomever he chooses. Moreover the negative device of preventing outsiders from occupying newly erected dwelling houses may not in the event serve the underlying purpose of ensuring the availability of adequate accommodation to local people. The effect of the control may merely be to increase the demand for older dwelling houses for use as holiday homes or for occupation by persons coming into the area to retire.
>
> (DoE 1983, 8)

The effects of this policy, which was implemented from 1977 to 1984 when it was amended in the light of the Secretary of State's deletion of the policy from the structure plan, were the subject of detailed analysis in Shucksmith (1981). That analysis is now briefly summarised to allow discussion of subsequent critiques and more recent empirical information on Lake District house prices during the period of the policy's implementation.

The effects of the LDSPB's policy

The board has explained (1978a, 20) that

> the policy arose from concern...at the growing difficulties which local people were having in buying houses in the National Park.

The aim of the policy, then, was to alleviate the problems of (young) local people in finding suitable accommodation to purchase.

By excluding commuters, retirement home buyers and second home purchasers from the market for new housing, demand was to some extent transferred from the market for new housing to the market for existing housing. Outsiders seeking to purchase homes in the national park were most unlikely to have been deterred by the new policy, but instead tended to compete with locals for the existing stock, over which the LDSPB had no control. Because of this diversion of demand from new housing developments to the existing stock, it is helpful to consider the effects of the policy on each separately.

The market for new housing

The effect of the policy on the market for new housing will have been substantial. The exclusion of non-locals from this market will have caused a shift in the demand curve. It is likely that this contraction was marked, since previously most new houses appear to have been bought by non-locals (see Table 6.2), and those local buyers able to afford a new house will almost certainly already have been owner-occupiers themselves, probably with no pressing reason to move house. Further, some difficulty was experienced by even these few locals in obtaining mortgages to buy houses subject to the section 52 agreement.

In addition, there may have been a leftward shift in the supply schedule as builders ceased speculative residential developments, partly because of the uncertainties raised by the new policy, but principally because of the greater difficulty of acquiring suitable building land with planning permission. Because only smaller developments were to be permitted, with consequently higher land prices, production costs would have been raised (G. Clark 1982, 104). The demand and supply shifts will have tended to offset one another in terms of the effect on price. Further, the supply of new housing is likely to have been sensitive to changes in expected prices (comparatively price elastic), after a time-lag (Charles 1977, 24). One would therefore expect there to have been relatively little reduction in the price of new housing, but the policy would be expected to have caused a large decline in the quantity of new housing bought and sold. This is consistent with objectives of restricting development and of retaining for local people the few remaining sites on which house-building is acceptable; but it is not likely to have contributed in any way towards meeting the needs of local young people, which were the *raison d'être* of the policy, according to Capstick (1987, 136). Indeed, Capstick (1987, 140) has confirmed that 'most of the houses built were not of the type which would have helped local first-time buyers to obtain homes', and therefore the policy 'did more to assist existing home owners...than it did to assist new home owners' (p. 144).

The market for existing housing

As non-locals transferred their attention from new houses to the existing housing stock there will have been an increase in the demand for that stock (a shift of the demand curve). The result of this increased demand is that there will have been an increase in the price of the existing housing stock, as more buyers sought these houses.

The size of the price increase depends upon the sensitivity of supply to changes in price (the price elasticity of supply): and normally one would expect the supply of houses already built and occupied to be highly insensitive to changes in price. The supply of existing dwellings coming on to the market depends less on price changes than on households' mobility, since households tend to move in response to change in their family or employment circumstances, or to a change in income, and tend not to move in response to changes in house prices. The supply of existing dwellings is therefore likely to be very price inelastic in the medium or short term, and so one would expect the price of existing houses in the Lake District to have risen substantially on a once-for-all basis as a result of the increase in demand for existing houses.

This substantial increase in the price of the existing housing in the Lake District as a direct result of the LDSPB's policy may have deterred some outsiders from purchasing second homes, commuter homes or retirement homes within the park. These people may have turned their attention instead to other areas, possibly including areas immediately adjacent to the national park, as South Lakeland District Council has claimed (SLDC 1980a, 5), so forcing up house prices there as well. Notwithstanding Capstick's (1987,137) view that this is 'inconceivable, since South Lakeland has its own high level of demand' (a curious argument when her own estimates suggest that Lake District prices are 20–30 per cent higher than those outside), she nevertheless provides evidence of just such price rises in areas adjacent to the national park.[1] However, this boundary effect is not a central issue.

Effects on local people and disadvantaged groups

The effects on local people can be summarised as follows. Local people who could afford to buy new housing will have found prices roughly the same as before, once the shifts in the demand and supply schedules had worked through. The structure plan envisaged that new housing would be restricted to infill developments of a type and density appropriate to the character of the settlement, and these were bound to be expensive. Any local person who could afford to buy a new house was unlikely to have been experiencing housing disadvantage in the first place, and could certainly have found housing without the help of this policy. The board's concern, and (one imagines) its policy, were not aimed at them but at those locals with more pressing need of accommodation. As G. Clark (1982, 104) has noted,

> If the greatest effect of the policy in reducing house prices proves to be for houses costing over £40,000, it is not clear how valuable is the claim that local need is being met.

For the disadvantaged groups amongst permanent residents it was the effect upon the rented sector and upon the price of the cheaper existing housing stock which was relevant. The effect of the board's policy in increasing the price of the existing housing stock will have made it harder than ever for them to afford to buy accommodation in the Lake District. It was in the market for the cheaper existing housing that such disadvantaged groups competed to buy houses, if at all, and the effect of the LDSPB's policies will have been to disadvantage the local population further in their attempts to purchase accommodation within the park. Surprisingly, Capstick (1987, 136) has argued that this effect is unimportant:

> The Board's whole case...was that local young people could not compete in the existing housing market. It was of no consequence to them that further outside buyers would now enter that market.

If this is accepted, then the board's policies will merely have been irrelevant, rather than harmful, to the needs of such groups. However, it seems inevitable that one consequence of a rise in the price of existing houses will have been to exclude marginal local purchasers, and thus to add to the numbers of those who cannot afford house purchase and must look instead to the rented sector.

Therefore, one might expect there to be more people seeking rented housing. The consequences for the supply of rented accommodation are less clear: while the rise in the vacant possession price of existing houses, relative to private sector rents, would have added some further disincentive to private landlords to relet when a protected tenancy ends, the disincentive is so large already that the policy is unlikely to have had any significant effect. In the public sector the main effect is on the cost of land for new building, and it might be expected that local authorities and housing associations would have been able (all other things being equal) to have purchased land subject to section 52 agreements more cheaply than otherwise. However, if this was the real object of the policy, land costs could have been reduced still further for public sector housebuilding by limiting the definition of local needs to public sector development only. This course is considered further in the next chapter.

Empirical corroboration – house prices 1970–86

This assessment of the effects of the LDSPB's policy has relied purely on deductive analysis, rather than on empirical observation. Having deduced these expected consequences, an attempt was then made to gather empirical information on house prices before and after the policy's implementation, in order to check whether the effects deduced are consistent with observed price trends. The principal source is a data-set of house prices from 1970 to 1986, gathered from local newspapers. This was the first systematic time-series of house prices in the Lake District to be collected. A similar, but far smaller, data-set has recently been collected by Capstick (1987), and this is presented for comparison.

Data were collected from the property columns of the *Westmorland Gazette*, a weekly newspaper which covers the most populous area of the Lake District, the southeast. For January, May and September of each year, from 1970 to 1980, all houses advertised for sale within the national park, for which an asking price was given, were included in the sample. The total sample collected in this way consisted of 799 observations. The exact means by which this data was analysed is described in Shucksmith (1981). The outcome of this analysis is a mean price for each type of house for each year, holding size of house constant. From these, a weighted mean house price was computed for each year, using weights for each type of house according to the overall proportion of each type in the sample. This weighted average house price has the virtue of holding both the size and type of house constant from year to year. It is thought that this sample, and therefore the price estimates, are representative of house sales in the south-eastern Lake District, with the exception of the tiny proportion of very expensive houses which are sold by auction. Recently, this sample has been supplemented by further data for 1981 to 1986, increasing the total sample size to 1,261 houses.

The estimates of house prices are shown in Table 6.5 and Figure 6.1. It should be noted that these are newspaper asking prices for the most expensive area of the national park, and on both these grounds they may be overestimates of house prices in the Lake District as a whole. These estimated house prices for the Lake District may be compared with national and regional house prices, as in Figure 6.1.

These results confirm that houses in the southeast Lake District are more expensive, on average, than those elsewhere. This accords with prior expectations, not only because the data for the southeast area of the Lake District refer to asking prices rather than to selling prices, but also because of the area's attractiveness.

More revealing, perhaps, are the rates of house price increase shown in Table 6.6 and Figure 6.2. These demonstrate that, although Lake District houses are more expensive, for most of the 1970s prices had risen little faster than those elsewhere. Between 1970 and 1972, and between 1973 and 1977, house prices in the Lake District increased at a slower rate than national house prices. In the year 1972–3, on the other hand, house prices in the Lake District leapt by 77 per cent.

Partly, this leap may be explained by a tendency of sellers to ask over-optimistic prices, as reports of the national property boom gained prominence: this hypothesis is supported by the fact that Lake District house prices actually fell in the following year. Also, Simpson (1975,

Table 6.5 Estimated average house prices in southeast Lake District, 1970–86

	2-bed terraced	*3-bed semi*	*3-bed bungalow*	*3-bed detached*	*Weighted average*
1970	3,206	4,372	8,281	6,738	5,807
1971	2,955	6,532	11,463	10,305	7,730
1972	3,992	5,567	11,918	12,237	8,864
1973	9,462	13,617	17,813	18,244	15,341
1974	8,878	10,197	20,541	24,275	14,852
1975	10,982	12,735	19,177	24,275	17,276
1976	11,739	14,670	21,650	23,814	19,431
1977	11,633	17,666	25,972	23,721	19,591
1978	13,966	21,470	30,178	33,868	24,902
1979	25,064	38,717	41,193	49,108	37,836
1980	28,592	37,101	47,449	51,554	41,234
1981	32,750	36,955	48,177	57,774	46,394
1982	29,066	36,849	49,928	60,045	45,678
1983	32,310	37,495	49,677	61,255	49,541
1984	32,724	39,092	59,636	66,325	50,370
1985	37,647	46,116	60,559	72,685	55,550
1986	43,111	47,154	58,326	63,554	53,321
Weight	0.34	0.10	0.26	0.30	1.00

30) found that in 1972 there was a doubling in applications for improvement grants in the Lake District, partly because of temporarily higher grants introduced in assisted areas in 1971 for home improvements which could be completed by June 1974, but also perhaps because 1972 was the peak of the growth in numbers of second homes in Britain (Shucksmith 1983, 175). This might account for the additional local increase in house prices, on a once and for all basis, since the incentives of improvement grants and tax relief for second home owners were removed in 1974, and the subsequent decline in second home numbers in Britain was not arrested until 1976–7.

Figure 6.1 Estimated average house prices in southeast Lake District, compared with regional and national house prices

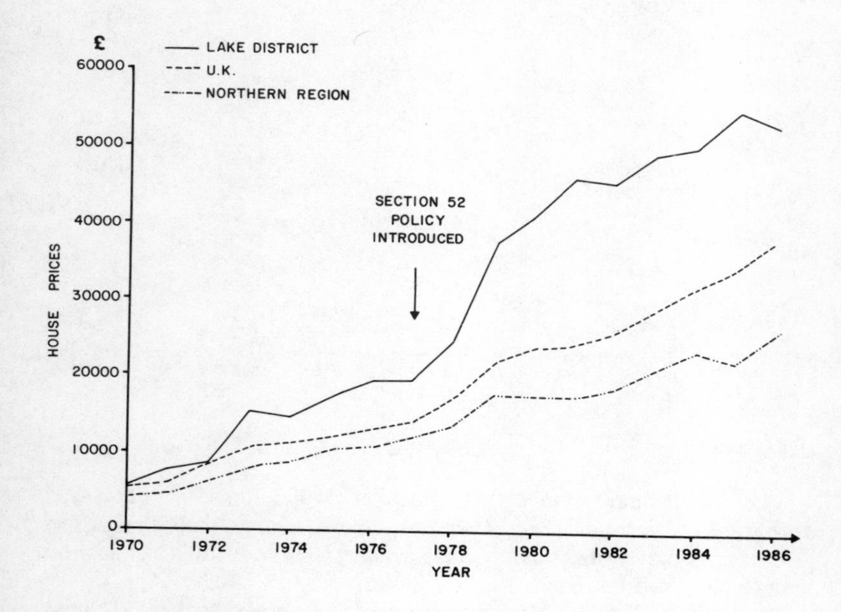

Sources: UK and North Region: DoE (final quarter). Southeast Lake District: author's sample.

Table 6.6 Percentage price increases, 1970–85 (percentages)

	UK	*SE Lake District*
1970–2	+65	+50
1972–3	+27	+77
1973–7	+32	+28
1977–81	+73	+137
1981–5	+41	+20

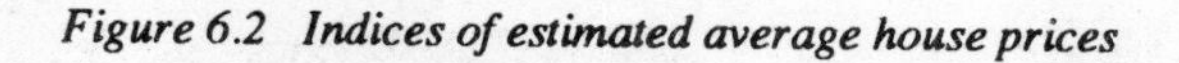

Figure 6.2 Indices of estimated average house prices

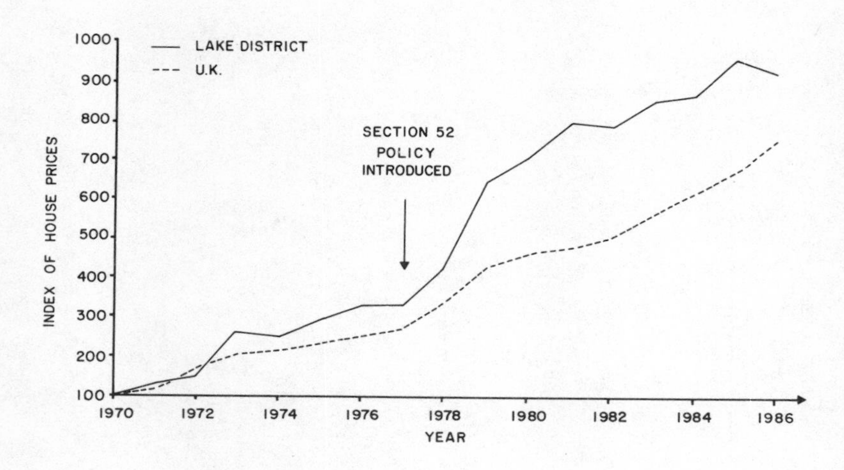

The rate of increase then slowed between 1973 and 1977, yet it was during this period that several suggestions were made that local house prices were increasing faster than elsewhere: such assertions can now be seen to be unsubstantiated.

Between 1978 and 1981 house prices rose sharply. In 1977–8 local prices rose by 27 per cent which was a similar rate of increase to the national figure of 22 per cent. In 1978–9, however, local prices appear to have risen dramatically, by 52 per cent, in comparison with a national increase of only 29 per cent. In 1979–80 prices both locally and nationally tended to be more stable, owing to the shortage of mortgage funds. Nevertheless, house prices in the Lake District were still increasing faster than the national rate. In 1980–1 the gap widened again as Lake District prices rose by 13 per cent compared to a national increase of 3 per cent. The total effect of these price increases was that by 1981 houses in the Lake District cost well over twice the 1977 price, whereas UK house prices had increased by only 73 per cent over this period. Figures 6.1, 6.2 and 6.3 illustrate the extent of this difference.

Since 1981, Lake District prices have not risen as fast as national house prices in percentage terms, although in absolute terms a differential of about £20,000 has been maintained (ignoring what may be an aberrant price estimate for 1986). Since 1985 it is notable that there appears to have been some divergence within the Lake District between rapidly escalating prices of cheaper types of housing and falling prices of the larger houses and bungalows: the prices of terraced and semi-detached houses have risen by 32 per cent and 21 per cent respectively, while those of bungalows and detached houses have fallen by 2 per cent and 4 per cent respectively. This probably results from a tendency towards sale at auction of more expensive houses. Leaving aside 1986, these estimates show overall that Lake District prices since 1973 have maintained a fairly constant differential over UK prices, apart from the period 1977–81 during which the differential of about £5,000 which characterised the mid-1970s jumped to the £20,000 gap of the 1980s.

The analysis presented earlier in this chapter predicted that the major consequence of the LDSPB's policy would be a substantial increase in house prices within and around the national park. The board's policy was introduced in Autumn 1977. Information from estate agents suggests that prices started to accelerate in the Lake District in November 1977; and the analysis of house prices presented here shows that the greatest increases appear to have taken place during 1978–9, in line with prior expectations of a lagged response to the board's policy. Although there were outstanding planning permissions in 1977, and various other lags in the system, the policy immediately created the expectation of future house price increases: this may have contributed to the early rise in prices which followed, perhaps sooner than might have been expected.

For the examination in public of the structure plan, the LDSPB undertook a similar analysis, and obtained similar results. The board disagreed about the causes, however, concluding that:

> If a reason for sharp increases in Lake District prices in the last few years can be determined, it is more likely to be a consequence of the dramatic reduction in the number of housing approvals given compared with earlier in the decade... Even if all permissions granted since 1977 had been free from local occupancy restriction, the increasing scarcity value of new houses relative to a few years ago would probably have been reflected in price increases much the same as have been witnessed.
>
> (LDSPB 1980b, 11)

This is an important point. The LDSPB itself attributes this reduction in the number of housing approvals to the success of the section 52 policy, congratulating itself that:

> So often new policies sound fine in print but have little impact on the ground; but here the results are being seen quickly.
>
> (LDSPB 1978a, 22)

Crucially, the reduction in housing approvals was conceived jointly, not separately, from the policy of imposing a local occupancy restriction: if the LDSPB had reduced housing approvals in some other way the effects would have been similar, as the board suggests, but it will be argued in the next section that the 'homes for locals' policy was as much a device for reducing development as for helping to meet housing needs, and that it is hard to think of any alternative means of restricting development which would have been politically acceptable.

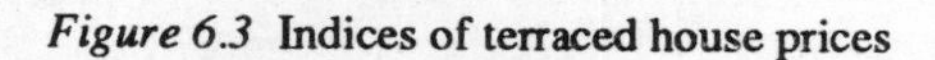

Figure 6.3 Indices of terraced house prices

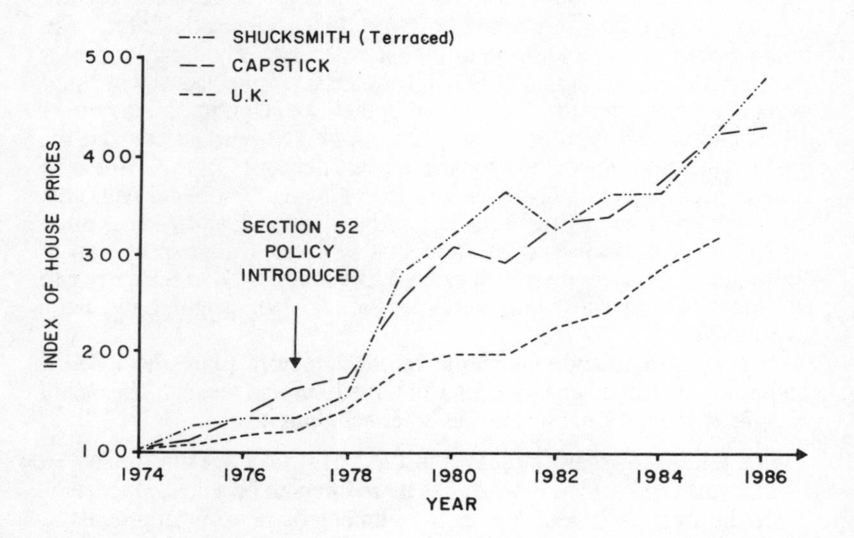

Note: Each sample is drawn from newspaper adverts. Sample sizes are 4.5pa. (Capstick) and 25.3pa. (Shucksmith).

More recent price estimates by Capstick (1987), shown in Figure 6.3, also tend to confirm the trends suggested. Capstick's (1987, 137) conclusion that 'Lake District prices rose broadly in line with national trends' appears difficult to reconcile with her published price estimates. It is clear from Figure 6.3 that her series of house prices, despite the much smaller sample, also shows a steep increase relative to UK prices between 1978 and 1980, and that it follows fairly closely the estimates of terraced house prices in Table 6.6. The deductive analysis of the effects of the LDSPB's policy suggested that one would expect a substantial increase in house prices to follow as a consequence of the policy. The empirical finding of a large increase in Lake District house prices following the policy's introduction during 1978–81 is certainly consistent with this expectation. There are other factors which might wholly or partly explain the price rise; such as the renewed growth in second home numbers in Britain from 1976–7, or perhaps an ability of retirement purchasers to buy a home in the Lake District without requiring a mortgage from building societies which at that time were short of funds. Also, the expected time-lag before prices rose appears to have been rather shorter than anticipated. In any event, causality cannot be demonstrated merely by observing price trends and this has not been the intention. What has been demonstrated is that Lake District prices tended to rise faster than national prices after the policy's introduction, and this tends, if anything, to corroborate rather than to contradict the *a priori* analysis. This is consistent with the conclusion that the board's policy led to higher house prices than would have been the case had the more relaxed planning policies of the early 1970s continued.

Two critiques

While the LDSPB has not challenged the substance of these arguments since they were first sent to them in 1980, a serious critique has been offered by Loughlin (1984). Loughlin argues that the reduction in residential planning permissions results from the LDSPB's conservationist stance rather than from the section 52 policy. The effects attributed to this policy should, he argues, be blamed instead on the LDSPB's protective attitude towards the landscape. Quoting the above account of two logical errors in the LDSPB's argument, he argues as follows:

> Taking Shucksmith's second point first, while one may accept his arguments on flow determining price, he may nevertheless have overemphasised the impact of the policy. Annual rates of housing approvals in the Lake District were steadily decreasing throughout the seventies (if the years 1973 to 1977 are taken, for example, the average annual rate is around 380). What Shucksmith has failed to demonstrate, therefore, is that, were it not for the existence of the local user policy, annual approval rates would remain at around 450 per annum. It seems highly unlikely that this would be the case. When the Board undertook surveys in the late seventies as part of the structure plan process, for example, it identified sites for only 1,050 dwellings in the national park as a whole which could be used without having an unacceptable impact on the natural beauty of the park. Consequently, even if the local user policy were not adopted it seems highly likely that approval rates would have significantly decreased. Thus, it seems that Shucksmith's analysis is directed, not towards the local user policy, but rather at the Board's conservationist stance. It is this which, in the context of increasing demand, has been primarily responsible for

> the rise in house prices in the Lake District. That this has not been recognised by Shucksmith seems clear from his first point which assumes that it is the local user policy which has limited the number of houses being built. In fact this is not the case. Conservation policies are primarily responsible. This would also explain why Shucksmith was able to discern an increased rate of price rises immediately after the adoption of the local user policy.
>
> (Loughlin 1984, 99–100)

Loughlin is quite correct in arguing that, if approval rates had declined to the same extent in the absence of the section 52 policy, then the conse- quences of the decline in approval rates should be attributed to the LDSPB's general conservationist policies. But it is not clear that the decline in approval rates would have taken place to the same extent without the existence of the section 52 policy, nor is it appropriate to attempt to separate these two policies. The discussion of the genesis and evolution of the board's policy showed how the policy arose largely because of a reluctance to impose more restrictive controls on development unless local people could be spared the housing consequences: indeed Capstick (1987, 129) regards the policy 'as a corollary of its strict control of development'. The section 52 policy made possible the board's more restrictive approach because it appeared to allow a more restrictive policy also to help local people find houses. A number of the LDSPB's statements reveal this attitude. For example,

> As the concern for the difficulties of local people grew, there was also a growing realisation that the rates of housing development of the late 1960s could not be sustained through the 1970s without producing major environmental impact. A stricter attitude towards house building was necessary if the character of the national park was to be retained. The theme of stricter control of development was taken up and developed by the Sandford Committee and endorsed by the Government, and has become an accepted part of national park policy. In the Lake District, however, the Board felt that whilst stricter development control was justified it should be matched with a policy which would attempt to help with the particular difficulties of local people.
>
> (LDSPB 1980b, 1)

G. Clark (1982, 96) also notes an intertwining of the LDSPB's concerns for landscape and housing in his account of the policy's origins:

> As a result [of Circular 4/76], the members of the Board held a number of informal seminars and concluded that they needed to do something to help local people obtain housing and to preserve more rigorously the Lakeland landscape.

Capstick (1987) explains how the policy arose from public reaction to the board's preliminary intimations that it would adopt a more restrictive attitude to new housing developments. To retain the legitimation and the support of the park's population the board was unwilling merely to impose such restrictions without addressing local housing needs. Consequently,

> the Board took the view that its overriding duty to preserve the landscape of the National Park laid such severe constraints on its possible approvals of new sites for building that the acceptable building sites which remained should not be used up in the provision of further holiday or immigrant retirement homes, but should be, in effect, released over a longer period of time and in a way which would meet local need.
>
> (Capstick 1987, 3)

Loughlin is therefore incorrect to attempt to separate the local user policy from the board's conservationist stance: the policy both follows from the more restrictive stance and at the same time legitimises it, in that it is unlikely that the board would have felt able to adopt such a restrictively conservationist stance without demonstrating (both to the people of the park and to themselves) that the housing interests of local people would be protected. The two elements are therefore inseparable.

More recently, Capstick (1987) has offered a critique of the economic analysis discussed. She disputes the conclusion that the policy's effect will include a rise in the price of existing houses, arguing instead that the price would remain unchanged because 'demand would transfer equally to the older stock from incomers, and to the new stock from residents' (Capstick 1987, 143). This argument, however, is fallacious, resting on an implicit and untenable assumption that the supply of new housing is infinitely price inelastic. Capstick's argument in full is as follows:

> Of, say, 100 new houses built, a proportion would have been built for or bought by people who could have fulfilled the local occupancy condition...The remainder would have gone to newcomers. This latter group alone would transfer its demand to the existing housing stock, but an equal number of local people would be able then to buy the new, conditioned houses. These would not perhaps be first time buyers, though some might be. But their demand would be removed from the existing stock, and the net effect of the introduction of the local occupancy policy would have no marked effect on the prices of existing houses.
>
> (Capstick 1987, 137)

There is no reason, however, to suppose that demand would transfer equally to the new stock from residents. Exclusion of non-locals from the market for new housing, as before, constitutes a shift in the demand curve: only if the supply curve is vertical (i.e. infinitely price inelastic) will the quantity of new houses supplied and consumed remain unchanged, since any resulting increase in demand from local people will be a movement along the demand curve rather than another shift. If the supply of new houses is at all responsive to price changes, on the other hand, then the quantity supplied and consumed will fall and consequently demand will not transfer equally to the new stock from residents. Indeed, if as is likely the supply of new houses is price elastic, then there will be relatively few additional residents transferring from the existing stock to new houses, and the net effect will be an increase in the demand for existing houses as originally proposed.

Inequitable consequences

The policy has not only been criticised on the grounds that it does not achieve its aims. It is also inequitable. It discriminates in favour of those who already own houses, in that their houses will increase in value as a result of the policy. Conversely, the policy discriminates against any local person who aspires to own his own home, and in two ways. First, houses will become more expensive as a result of the policy, and first-time buyers will find house purchase even more difficult; second, the purchasers of new houses covered by the section 52 agreement have found difficulty in obtaining viable mortgages because of the restricted resale market.

Other groups are also affected. Owners of land with outstanding planning permission have benefited at the expense of others who might have hoped to obtain such permission. Local builders have complained that the policy threatens employment opportunities in an already contracting industry. Non-locals are the object of explicit discrimination which prevents them from building new houses within the national park, although their freedom to purchase an existing house is unaffected. The principal beneficiaries of the policy, apart from local owner occupiers, are recreationists, tourists, and all the visitors to the Lake District for whom the landscape and natural beauty is to be preserved and enhanced.

Conclusion – the impotence of local planning

These conclusions return us to one of the principal themes of this book: the conflict of public policy objectives. Only the existence of such a conflict can explain why the LDSPB has introduced and continued to promote a policy, ostensibly to help local people, which tends to inflate local house prices to the detriment of potential local purchasers, which is demonstrably regressive in its distributional effects, and which meets with ministerial disapproval.

The primary duty of the national park authority is to preserve and enhance the natural beauty of the national park. With this duty uppermost in their minds, the members of the planning board reached the conclusion that very few additional houses could be built in the national park without seriously affecting the landscape; but at the same time they were reluctant (perhaps even unwilling) to implement a highly restrictive development control policy without some compensating mechanism for helping local people find housing in the park. Capstick (1987, 2), a leading member of the LDSPB at the time, recalls that

> When, in 1974, the newly established Lake District Special Planning Board began work on the National Park Plan, it became apparent at each public consultation meeting, and in written responses, that the chief matter of concern for people in the Park was the inability of local workers to buy or rent houses. The Board had seen the result of the well-meant efforts of its predecessor to solve this problem by allowing the building of more houses, and did not like what it saw.

The problem for the board was that any attempt to increase housing provision in the national park would, they believed, conflict with their primary duty of conservation and preservation. There was no way in which the occupancy of existing houses could be altered in favour of local people, as far as the board was aware. The frustration of the board's members was apparent (Shucksmith 1981).

Essentially, when the board was faced with these two conflicting objectives, any reconciliation of local interests with the duty of landscape preservation appeared impossible. Capstick (1977, 515) emphasised this point:

> The present Board...has realised that its most daunting task is not the reconciliation of landscape preservation and public access, difficult though this is, but the retention of a living rural community in the face of pressures from wealthy urban interests, and under the entirely necessary constraints of the need for landscape preservation.'

The board appeared to have reached an impasse.

Then, the suggestion was put forward that, even with a restrictive control of new housing development, local interests would be served by a policy which permitted new housing only where those houses would be occupied in future solely by locally employed people. It was immediately apparent to the board members that this policy would equally serve the objective of landscape preservation. At last, relieved members of the planning board thought that they saw an escape from the impasse of conflicting objectives. Here was a policy which appeared to reconcile local interests with the conservationist priorities of the national park. It was too good an opportunity to miss:

> new housing would only be permitted if it could be seen to benefit the local communities ..., (and) moreover, the Board intended to pursue the idea of somehow ensuring that any new houses would be limited to occupation by local people.
>
> (LDSPB 1978a, 21)

The section 52 policy resulted, and the board has clung to the idea ever since, even though the Secretary of State's deletion of the policy in 1984 has prevented its implementation since then.[2] Thus, it was reported in June 1987 that the LDSPB was discussing a series of recommendations from its officers in order to arrive at a new housing policy:

> The first of these is that the Board should ultimately aim to ensure that new housing is subject to a local occupancy condition... 'requiring all applicants for new residential development to justify their need for it'. Local occupancy restrictions were ruled out by the Department of the Environment's ruling on the joint structure plan. But Lake District planners are looking at the notion again...
>
> (*Planning* 724, 3)

Given this historical perspective of impasse and frustration, one can understand the reluctance of planning board members and officials to concede that the section 52 policy is not, after all, an escape from that impasse. The relief and hope vested in the introduction of the policy have given it considerable momentum, and it must be very difficult for its authors to realise that in fact it does not act to reconcile local interests with the principal objective of the national park, but instead tends to worsen and reinforce the housing problems of local people.

The 1978–9 Annual Report of the LDSPB betrays not only the reluctance of the board to abandon its progeny but also its bewilderment:

> It was certainly noticeable at public meetings and in press comments that, whereas before the adoption of the policy there had been criticism that the Board were doing nothing to help local people, now that criticism has died. It might be replaced by comment that their efforts were misguided, but at least it was generally accepted that the Board were doing something. It is a moot point as to which is the less uncomfortable position: being criticised for doing nothing, or being criticised for doing something but doing it wrongly... It will be understood from all this that

> the debate continues, and that Board members are very much aware of the different views expressed by many people. Nevertheless the Board feels that the policy is right. With experience and time there may be modifications to it, but the central aim remains. (LDSPB 1979, 13–14)

It has been suggested in this chapter that the reason why the policy was not abandoned is largely that the members of the board could not bring themselves to accept the criticisms made of the policy, whatever their merits: for to accept the reality of the policy's effects would have meant a return to the previous impasse. The reality of the situation was that no reconciliation of local interests and landscape preservation was achieved, and that there was no alternative mechanism open to the board through which these objectives could have been reconciled, given the legal and financial constraints under which it operated. As Shaw (1980, 90) has observed:

> the use of Section 52 agreements by the LDSPB to restrict permissions to applicants who work 'locally' (from October 1977) is significant in symbolising the frustration of planning authorities who can see the problem clearly but lack the policy levers to achieve its solution.

In this instance, central government has given the local planning authority a complex and inherently conflicting set of objectives, but denies it the means to resolve this conflict. Alternative responses might succeed in both protecting the landscape and relieving housing difficulties if some of these constraints were relaxed by central government; for example, by allowing local housing authorities sufficient finance to provide council houses and by allowing planning authorities to discriminate between local housing authorities and other residential developers in granting planning permission. These issues are the subject of detailed discussion in the next chapter.

Notes:

1 Capstick's evidence does not, though, substantiate her claim that 'house prices in the popular southern part of the Lake District actually rose less fast than did the prices of comparable houses in Kendal' (1987, 137): the position alters according to the year chosen as base.

2 Indeed, the LDSPB's 1988–9 Annual Report says that the board will exert pressure on the minister 'to allow a change in policy which would permit planning authorities to use agreements with developers to ensure that suitable building land could be made available for local people. Such an approach had been attempted previously by the Board but was rejected by the Secretary of State.'

Chapter seven

The local state: options for intervention

The discussion in the first five chapters, together with the findings of the case study of the Lake District, confirms that a conflict of public policy objectives exists (to a greater or lesser degree in different areas) between the release of land for housebuilding in rural areas and the protection of land from development in the interests of landscape and wildlife conservation and farmland protection. While in some areas policies may have been over-protective, nevertheless government must mediate between these competing claims, or seek to find means of meeting housing needs without detracting from the landscape. At the same time, and related to this, it has been clearly demonstrated that certain social groups are particularly disadvantaged in the competition for housing in rural areas, with young people and lower-income groups having especially limited housing opportunities. To a large extent this stems from a shortage of houses to rent.

The next two chapters consider what policy responses might be employed to adjust the balance between these competing claims on rural land, while also seeking to improve the housing opportunities of presently disadvantaged groups. In particular, how might government encourage the provision of low-cost housing without damaging the landscape? Such a question reveals the virtual impotence of local planning authorities, and strikes at the fundamental assumptions of the British planning system. It also necessitates a detailed analysis of the financial and administrative constraints which limit the contribution of local housing authorities.

In addition other potential policy responses are considered, both under the existing constraints described and, more radically, in the absence of these constraints. These range from current initiatives in Britain (such as short-leasing schemes and housing association activity) to policies which would require some enabling action on the part of central government (such as relaxing or modifying advice on planning controls, municipalisation and nominated leasing and land settlement).

Each of these possible mechanisms is assessed both from the point of view of economic efficiency and from an equity perspective. In Chapter 8, these options are viewed from a central strategic perspective, but this chapter focuses on the mechanisms available to the local state.

Two principal justifications for public intervention have been proposed. In Chapter 3 it was argued that the economic and social welfare of rural communities is a form of producer good, essential to the production of two public goods, landscape preservation and public enjoyment. Therefore, while the control of new housing development may be desirable in order to protect landscape on efficiency grounds, the provision of housing which helps to sustain the rural community is equally necessary to the long-term objectives of countryside policy. These public good aspects, and the inherent conflict apparent between these two aims, provide one rationale for public intervention in the provision and control of rural housing.

The other justification is on grounds of equity. In Chapter 4 it was noted that housing is often viewed as a merit good, and indeed that successive governments have espoused the aim of 'decent homes for all' at a price within their means. Yet Chapter 5 presented evidence to suggest that some groups within rural society do not have access to a decent home at a price or rent within their means, and that clear patterns of disadvantage exist. Elderly people and the unmarried under thirty, without secure accommodation or a good income, have the most limited range of rural housing opportunities. Assisting these groups to improve their housing opportunities is a further rationale for public intervention.

The case study has both reinforced and qualified these concerns. In the Lake District the public good aspects of landscape and rural housing, and the inherent conflict between aims of house prevention and provision, are particularly prominent, with apparently little room for compromise. As a consequence, in part, lower-income groups were severely disadvantaged in the competition for housing: as well as posing a threat to the future social and physical character of the area, it seemed that this was a most inequitable outcome. The policy response which had bravely attempted to reconcile 'local needs' with landscape preservation was unsuccessful in helping the disadvantaged groups, largely because planning mechanisms *on their own* are unable to allocate housing or housing land to particular social groups, but also because the detailed effects of the policy had not been foreseen.

Given these objectives of public policy, the structure of disadvantage which has been observed, and the shortcomings of policies in the case study area, it will be helpful to review a range of policy mechanisms which might be employed in pursuit of these objectives.

Planning measures

In the first place, the potential contribution of existing planning powers will be considered. In Chapter 3 the conclusions of Hall *et al.* (1973) concerning the regressive distributional effects of the operation of the post-war planning system were summarised, and it is necessary now to ask why this is so. Has planning necessarily favoured the better off at the expense of the poorer members of society, or is this merely a consequence of the way in which the planning system has been implemented ? Turning this question around, would it be possible to use existing planning powers to favour the disadvantaged members of rural society?

Hall *et al.* (1973) argue that the idealised planning system envisaged by the wartime committees (Barlow, Scott and Uthwatt) was corrupted in its later implementation: among other departures from these committees' intentions, the market in land was not abolished, and private enterprise rather than the public sector took the initiative. Uthwatt intended that all land on which development was permitted would first be acquired by the state at its existing use value and then leased to the users. Instead a distorted market ensued in which the conferment of development rights created private monopoly value in land. This 'crucial weakness of the post-war planning system' provoked a rapid inflation of land values, which in turn forced builders to economise on land by raising residential densities. As experience in Sweden demonstrates (Duncan 1983), restricting the supply of housing land need not have led to inflated house prices if the state had retained the development value as Uthwatt proposed.

Given the failure to recoup betterment, it was almost inevitable that the planning system should favour the better off. Hall *et al.* (1973) suggest that a shortage of building land was created in rural areas by local planning authorities, which resulted in rapid increases in house prices and a socially exclusive rural housing market. The shortage of building land was brought about for several reasons; foremost among them was the concern to protect agricultural land and the amenity of the countryside. Hall *et al.* (1973) recognise the power of the agricultural and rural preservationist interest groups in restricting rural building, but they also point to other forces.

> Basically, the argument was that planning powers, and the ideas underlying those powers, have been used to secure objectives which the planners themselves never envisaged. Protection of the physical character of suburban and rural areas has been used as a pretext for protecting their social exclusiveness ('apartheid'). The suburbs and rural counties, by resisting the outward spread of the cities, have been able to avoid the rate increases which influxes of

> working-class residents into their areas would otherwise have necessitated. Conversely, the cities have been able to retain both rateable values and Labour voters.
>
> (review of Hall *et al.* in Reade 1982, 79)

Whether restricting rural housing development is genuinely in the interests of landscape protection, or a mere pretext, the operation of private markets in land and in housing thus ensure social exclusiveness. If planning merely restricts rural housebuilding then, in the absence of Uthwatt's proposed abolition of the private land market, its consequences must be socially regressive, leading in time to the gentrification of the countryside and a fundamental change in the character of rural areas.

Often this appears to suit rural councillors very well, as argued by Newby (1980) and Hall *et al.*(1973). Rocke (1987) describes how Wansdyke District Council's favour of private ownership and minimal intervention has been assisted by the apparent separation of social from physical concerns in development control, allowing the council, for example, to draw tight housing development boundaries around villages. He makes the general observation that in rural areas

> investment priority is low and the voice of restraint loud. Hence, local planning and development control have become increasingly bound to the interests of 'established' residents and the restraint lobby in rural environments at the same time as the need has arisen for greater flexibility and discrimination between the interests of many competing social groups.
>
> (Rocke 1987, 182)

However, a growing number of rural councils have (ostensibly at least) been seeking alternative policy mechanisms with which to discriminate in favour of disadvantaged groups, such as young local people (Rogers 1985a; NACRT 1987). According to the LDSPB (1979) several other planning authorities were monitoring the progress of their 'locals only' policy with a view to implementing it themselves if it proved effective.

From this perspective, the weaknesses of the LDSPB's 'locals only' policy were twofold. In the first place, the private land market was not by-passed or replaced but merely sub-divided into two: within each segment (existing housing for all, new houses for locals only) the socially regressive effects of a severe restriction on supply remained. Second, and related to this, the policy attempted to discriminate on too crude a basis between locals and non-locals, rather than on the basis of housing disadvantage and need. What was really required was to replace the private land market and the private housing market with an alternative allocative mechanism which, unlike the private markets, favoured either those groups with the greatest housing needs or those groups with the most to contribute to national park purposes.

Local planning authorities may be able to make some progress towards that end, even under their existing planning powers. The Secretary of State's deletion of the 'locals only' policy from the structure plan was primarily on the ground that 'planning is concerned with the manner of the use of land, not the identity or merits of the occupiers' (DoE 1981a, 10–11). Yet this is a rather narrower view than that taken by the courts; for example, in the case of Fawcett Properties v Buckingham County Council, 1961, which accepted that a planning condition might properly be imposed to restrict the occupancy of certain dwellings to agricultural workers only, or in the case of Great Portland Estates v Westminster City Council, 1984, which accepted that development *plans* might properly take account of the land user's identity. The 1947 Town and Country Planning Act states that planning authorities may take into account 'other material considerations' when determining a planning application, and much legal opinion suggests that these considerations may include social and economic objectives of the planning authority where these are relevant to land use itself. Thus, a lengthy debate in the *Journal of Planning and Environmental Law* (1974, 309, 342, 410, 470, 507, 660, 718) followed the case of R v London Borough of Hillingdon *ex parte* Royco Homes Ltd, 1974, on the question of whether a planning authority could refuse permission for private houses to be built because it wished the site to be used for public housing. The court left this question unanswered, and opinions differed as to whether this was a proper planning ground. Thus, Brown (1974, 508–9) argued as follows:

> Conditions relating to occupancy of a dwelling-house have been held to be valid where the limitation has some planning purpose. Thus a condition may validly require a house to be occupied by a person employed or last employed in agriculture, because of a need to secure that agricultural workers shall be able to occupy suitable rural houses.... Nevertheless, the refusal of planning permission because the land is needed for municipal housing surely is unrelated to planning considerations and is improper.

Garner (1974, 511) disagreed, however:

> Many local residents would have strong views as to whether development in their vicinity was to be private or Council; however misplaced such views may be, they are surely not irrelevant in a planning context. Nor is it irrelevant, one would think, for the planning authority to take into account in making

> planning decisions such need as there may be for council housing in their neighbourhood.

This view was supported by Moore (1974, 513) who addressed the key question of whether the determination of priority between competing housing needs is a planning matter at all:

> In 1947...the answer to this question would surely have been that it was not. But planning today is recognised as an instrument of social change, a fact perhaps meekly borne out by the requirement that a local plan written statement must contain an indication of the regard the local planning authority have had to social policies and considerations.

This point is pursued by Young and Rowan-Robinson (1985, 223–4):

> The Structure and Local Plans Regulations specifically require planning authorities to have regard to social policies and considerations in the preparation of development plans and it would seem strange that they should be ignored for the purposes of development control.

Grant (1982, 338) takes this argument further by arguing that the formal inclusion of such a social policy in an approved local plan may itself be 'sufficient to accord it the stamp of materiality for planning purposes'. These textbooks therefore support the view that a planning authority may discriminate between public and private housing on planning grounds.

This power, if supported by the courts, has great potential significance for rural authorities facing a dilemma between restricting development and seeking to address housing needs of lower-income groups. Such an authority might argue in its local plan that the planning objective of maintaining the physical character of the landscape requires the retention of a local working community; and that the low paid occupations involved necessitate the provision of public housing to rent. On the other hand, private housebuilding is incompatible with landscape protection. Therefore a policy might be included to the effect that in order to maintain the physical character of the countryside all land suitable for residential development would be reserved for public housing. It could then refuse to allow land to be developed for private housing, giving the reason that the land is required for public housing, for the proper planning reason that this conforms to a policy in the approved local plan aimed ultimately at maintaining the physical character of the area. The argument employed is that developed earlier in Chapter 3, that housing for certain disadvantaged groups is necessary to the maintenance of the working rural community which supports the attractive landscape.

What is surprising is that until recently no local authority had deployed this argument and attempted to discriminate between public housing and private housing when determining planning applications in rural areas where private housing is socially exclusive. After all, such a course was recommended by the Sandford Committee as long ago as 1974. Sandford (HMSO 1974) argued that planning authorities in national parks, and other similar areas such as green belts, should adopt an even more restrictive policy of development control while housing authorities increased their provision of council housing in order to help the local people who would suffer as a result. The report argued that:

> The solution lies with the housing authorities rather than with the planning authority. Housing authorities can choose whom they will accommodate or assist and can thus, unlike the national park (planning) authority, discriminate in favour of local people. We are convinced that only in this way can the housing difficulties of local people be eased without destroying the essential long-term policy of restraint upon development.
>
> (HMSO 1974, 12.24–12.25)

Nevertheless, housing authorities could not achieve such an end on their own: Sandford's prescription relies implicitly upon the planning authority restricting private development, but allowing public sector housing to be built.

Interestingly, this is very much the approach recently adopted by the NAC Rural Trust (NACRT) in their attempts to encourage housing associations to build low cost homes for local people in English villages (NACRT 1987, 1989). The NACRT approach is instructive for a number of reasons, but principally its attraction is that it fulfils the two necessary conditions for meeting the needs of disadvantaged groups in rural areas: these are (1) the building of low cost housing in rural areas, affordable by those with the least means; and (2) the allocation of that housing in such a way that it goes to meet the needs of such households in the short- and long-term. It might be noted in passing that earlier 'local needs' policies, such as that applied in the Lake District, were deficient in both these crucial respects. NACRT achieves its success both by reducing the costs of provision through the acquisition of cheap land, and by allocating houses through non-market mechanisms.

According to Constable (1988), land costs are a major obstacle to the construction of low cost housing in English villages. Land typically accounts for perhaps 40–50 per cent of the total costs of provision if the open market land value is paid, and in some areas this proportion is even greater. Clark (1988, 59) reports that

> Land is the key factor in the affordable housing equation. In much

> of rural England, particularly in the south-east, any land with outline planning consent will fetch as much as £500,000 an acre (even £750,000 in some areas), compared with £700 to £2,000 an acre as farmland. The minimum cost for just the building plot is greater than...£51,000 in the south-east.

In order to reduce costs, and thus to pass on savings to the eventual consumers, NACRT has developed a method of acquiring land cheaply, at a price well below the market value with residential planning permission. NACRT seeks sites peripheral to the village and hence, not zoned in local plans for residential development. Since it cannot afford to acquire land at its full development value, NACRT is seeking land which is not scheduled for building and which therefore does not have full development value attached to it. If the landowner is willing to sell the land to NACRT, and the parish council approves, NACRT approaches the planning authority asking it to make an exception to its normal policy of development control and to give permission for a rural housing association to develop the site for low cost housing for specified (local) groups. Frequently, either the landowner attaches a restrictive covenant to the land or the landowner, the developer and planning authority sign a section 52 agreement to define these terms, and effectively to restrict the planning consent to the provision of social housing. Several local authorities are cooperating with NACRT in this way to reduce the land cost element in the supply of social housing, which is then allocated by the housing association according to need rather than ability to pay, whether for rent or for shared ownership.

This policy avoids the principal weaknesses of the LDSPB's earlier 'locals only' policy. New housing will not now be allocated by the market to more prosperous locals, but by local authorities and housing associations to those in most need of housing (presumably local or not, according to their published allocation criteria). This is not only progressive rather than regressive in its effect, but it is also far more sensitive as a mechanism than the blunt approach of a local occupancy condition. While the price of existing houses will still be inflated if such a policy is combined with the virtual cessation of new private development, this is less important to disadvantaged groups than the securing of a supply of rented housing allocated according to need. If successfully applied in areas of restrictive development control, such as the Lake District, this would lead not to a two-tier market in housing but to the tenurial polarisation of the rural population, with the wealthy occupying private dwellings and other groups in housing association and council dwellings. While this might not be ideal, the alternatives would seem to leave poorer groups without any form of housing at all. In less pressured areas, such low cost schemes might proceed alongside new private developments, without a concomitant inflation of house prices (Middleton 1988; Shucksmith and Watkins 1988e).

The success of such a policy depends upon two factors outside the control of the planning authority, however. In the first place, while the policy's inclusion in the local plan is not subject to ministerial approval, unless called in, he has the opportunity to overturn permission refusals for private development on appeal. Thus, in 1987, Ashford Borough Council included such a policy in a local plan which was approved despite objections from the DoE that the policy is unacceptable because it benefits particular groups and interferes with the disposal of private houses. The council argued that development was generally undesirable, but that exceptions should be made to allow local housing associations, for example, to build to meet local housing needs, and it cited the case of Great Portland Estates. While the inquiry inspector accepted this argument, recommending only that specific sites should not be shown in the plan, due to their exceptional nature, it would still have been open to the DoE to overturn refusals of planning consent on appeal, and so to breach the policy.

In the New Forest, however, not only has such a policy been included in the local plan: both the local plan itself and subsequent planning decisions have been successfully defended on appeal. In many respects the New Forest approach has been presented as a model for other authorities. For this reason it is helpful to summarise that approach (relying on an account by Croft 1988).

Because of the New Forest's AONB status, the district operates a very restrictive development control policy. But because of its concern for affordable housing to be provided to meet local needs, its local plan provides exceptionally for limited small-scale development to meet a particularly identified local need which cannot be accommodated in any other way. Schemes must be managed in the long term by, for example, a housing association or a village trust. Thus, the Forest and Downlands Villages' local plan states that

> Exceptionally, and subject to [other policy statements], the Local Planning Authority may be prepared to permit residential development which can be demonstrated to meet a particular local need that cannot be accommodated in any other way. In or adjoining villages where there are adequate local facilities (e.g. schools, shops and public transport) residential schemes whose occupation can be controlled in the long term may be permitted. To be considered favourably any proposal must be demonstrated to be economically viable; and to be capable of proper management, by for example a village trust or similar local organisation. Proposals to construct dwellings offering a discounted initial

purchase price only will not normally be considered to be within this policy.

According to NACRT (1988), New Forest District Council worked hard to build up a detailed picture of local needs in their villages, because they knew that on appeal against this policy they would have to demonstrate 'identified local need' existed. In the event, they have been successful in defending the policy against private developers on appeal. Clearly this is crucial to the success of such an approach, since only by establishing that such land will not be given full approval for private housing can landowners of land unidentified for development be persuaded that their land has only agricultural value: if landowners perceived any prospect of gaining full development value then they would be likely to require the full price from any housing association or village trust which approached them with affordable housing in mind, and the low cost element in the scheme would be fatally compromised.

In the New Forest, when approached by a suitable organisation such as a housing association or village trust, the planning authority considers giving consent, subject to an agreement being drawn up between itself and the developer or landowner, stating that the development must thereafter always be let and managed by a housing association in accordance with its charitable objectives; and that the housing association will try and let to local residents or people with a strong local connection. This agreement (a section 52 agreement) is deliberately loosely worded to allow for flexibility, and because the scheme's success ultimately depends not on watertight legal conditions but on goodwill between the authority and the housing association. So far, New Forest District Council has been fortunate in finding a number of landowners who are prepared to accept prices close to the agricultural value for their farmland, so allowing a number of low cost schemes to proceed (NACRT 1988, 1989). Most importantly, the policy has not only been successfuly defended upon appeal, but appears to have won the endorsement of the Secretary of State.

Neither the DoE's (1988b) discussion paper nor the Secretary of State's accompanying statement on housing in rural areas referred to these methods of acquiring cheap land, although they gave general support to NACRT's work. However, during the course of the Richmond by-election campaign in February 1989, the Secretary of State announced that planning advice was being revised to encourage the building of subsidised rented homes or low cost homes to buy in rural communities, along the lines of the NACRT approach. Sites which would not normally be released for housing development may exceptionally be released for low cost schemes if the planning authority is satisfied that there is a need for such housing, and that arrangements will be made to reserve it for local people (*Hansard*, 3 February 1989).

> In such areas, the need for low-cost housing and the existence of arrangements made by the developer or between the developer and the landowner or the local authority, to ensure that new low-cost housing is made available for local needs could be material considerations which the authority would take into account in deciding whether to grant planning permission.... Since planning conditions cannot normally be used to impose restrictions on tenure or occupancy, the planning authority would need to satisfy itself before granting planning permission that other secure arrangements to that effect would be made. Examples of such arrangements might be the involvement of a village trust or housing association with a suitable letting policy; covenants designed to give priority to first-time buyers from the locality; or an agreement between the planning authority and the developer under s.52 of the Town and Country Planning Act 1971. It would be important for schemes to ensure that the benefits of low-cost provision pass not only to the initial occupants but to subsequent occupants as well.

The announcement also stated that local plan policies should make clear that the release of such sites was in addition to the adequate provision of sites to meet general housing demand: and land allocated in the local plan to meet general housing demand could not be confined to local needs only.

Further encouragement was given in the 1989 Budget, when the Chancellor announced tax concessions to landowners who offer cheap land to rural housing associations as part of this approach.

The government has thus endorsed, and even facilitated, the use of planning measures to help reduce the cost of the land element in the provision of social housing in rural areas, and has stressed the importance of by-passing the market mechanism in the allocation of such housing. Its reasons for doing this, after opposing such schemes for several years, are discussed in Chapter 8.

Yet even this is not sufficient to ensure the success of such an approach. The other necessary factor upon which the success of such schemes depends is the willingness and the ability of local housing associations or the local authority to build social housing on the sites identified. Housing associations still construct only a very small number of units in rural areas, and the reasons for this are discussed on pages 157–63. Similarly, there is little council house building in rural areas, and the reasons for this are discussed in the next section. The general lack of will to build houses on the part of pre-reorganisation rural councils has already been described in Chapter 5, and no doubt it applies also to some authorities in rural areas today. However, many rural authorities have sought to build council houses in greater numbers since reorganisation, only to be prevented by centrally imposed financial and administrative constraints.

Constraints on public housing provision

While much of the shortage of rented housing in rural areas derives from the former local authorities' neglect of rural housing problems, the attempts of the new local authorities to make amends have coincided with a devastating reduction in investment in public housing, imposed by central government, together with a radical restructuring of housing subsidies which has systematically disadvantaged rural housing authorities. In order to explain these processes it is necessary to review in outline the system of finance for local authority housebuilding. The details of this system differ between Scotland and England and Wales, but their fundamentals are similar. Therefore, to avoid confusion it is the Scottish system which is outlined here to illustrate these constraints.

To build a council house a local authority must borrow money from the financial markets to meet the capital cost. Annual repayments, with interest, are then made over a term of up to sixty years from the current account of the authority. Capital spending therefore, has implications for the authority's current account, which must balance its annual expenditure, on loan repayments primarily, with its annual income from rents and other sources. Figure 7.1 illustrates the interrelationships between capital spending and recurrent spending (the housing revenue account, or HRA), highlighting the four crucial points at which central government exerts control over the local housing authority. These are controls over capital borrowing [1], the housing support grant (HSG), [2], the rate fund contribution (RFC) [3] and the rate support grant and the rate poundage itself, [4]. Each means of control is described in the course of the following discussion.

The greatest control over the scale of council building is exercised through the capital allocations announced each year by the Secretary of State. Capital allocations provide no funds for housing provision: they merely represent a permission to borrow. Councils are not allowed to borrow money for housebuilding or refurbishment beyond the sum announced by the Secretary of State, and in this way central government is able to limit the level of housing investment very effectively. Since 1976, successive ministers have frustrated the nascent enthusiasm of the new local authorities to build rural council houses, because of the increasing policy emphasis given to restraining local authorities' expenditure, particularly on housing. From the Labour government's approach to the IMF in 1976, and especially since the election of a Conservative government in 1979, the overriding imperative has been to cut the public expenditure devoted to housing, as a consequence of the monetarist macroeconomic policy pursued. This necessity has been combined under the Conservatives with two objectives derived from an ideological commitment to *laissez faire*, namely the reduction in size of the public sector and the increase in council house rents to unsubsidised levels (Gillett 1983). The effect on rural authorities of this withdrawal of subsidies is considered later.

Figure 7.1 Financial controls over local government housing

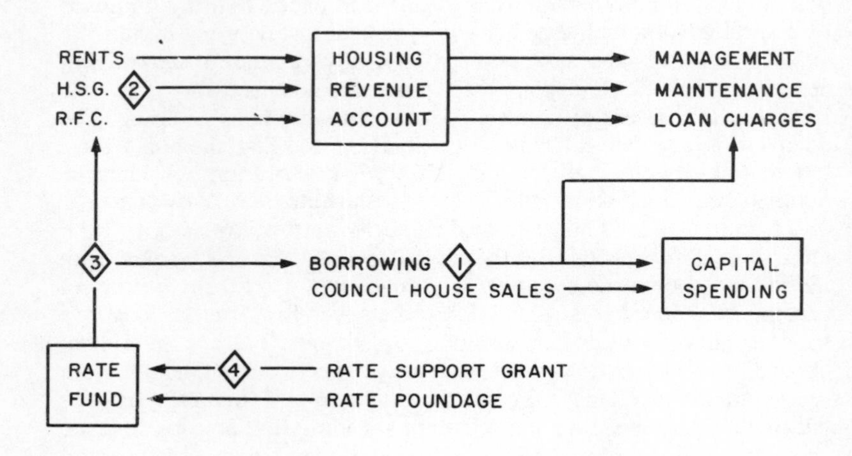

Table 7.1 shows the extent to which capital allocations and hence, capital expenditure on housing have been reduced for local authorities in rural and remote areas of Scotland. In Scotland as a whole, the level of investment in council housing has been reduced by more than a half in real terms over the last decade: from £1,198m in 1974–5 to £519m in 1985–6 in constant 1983–4 prices (*Hansard* 11 November 1985, col. 63). Rural areas have suffered similarly: between 1976 and 1986 real investment in council housing fell by 68 per cent in rural and remote Scotland. In England and Wales there has been a similar real reduction of 69 per cent in capital allocations since 1979–80 (*Hansard* 4 February 1987, col. 718). In fact, Table 7.1 understates the reduction in Scottish capital allocations by presenting gross capital consents: since 1981 permitted borrowing has been reduced by the notional capital receipts which authorities are deemed to have available from council house sales, whether or not such a level of sales is realised.

Table 7.1 Capital investment in council housing in rural and remote Scotland (constant 1985–6 prices)

	Capital allocations
1975–6	(£257.1m)[a]
1978–9	£91.9m
1979–80	£89.4m[b]
1980–1	£77.5m
1981–2	£84.2m
1982–3	£93.6m
1983–4	£97.9m[b]
1984–5	£75.1m
1985–6	£80.0m
1986–7	£87.7m
1987–8	£92.6m[b]

Notes: Figures relate to HRA spending, and include capital receipts from council house sales.
[a]The 1975–6 figure is capital expenditure since this pre-dates capital allocations.
[b]indicates years of general elections (peaks).
Source: calculated from CIPFA Rating Reviews

Neither do receipts from council house sales represent an encouraging alternative source of capital for investment in new building. Most research on the subject, except the Treasury's, has found that local authorities lose financially as a result of council house sales at current discounts (English and Martin 1983). Capital receipts must be weighed against the future loss of rental income, and against the costs of building houses to replace the lost relets. In rural and remote Scotland, where the differential is greater between the price obtained for a house at the district valuer's valuation less discount and the very high costs of replacement, several houses must be sold to finance the construction of one replacement to the stock.

Rogers (1981, 162–3) has reviewed capital allocations to rural authorities in England and Wales, and demonstrated not only that these were proportionately only about half the national average (being £34.65 per capita compared with £61.7 per capita nationally in 1979–80), but also that the more rural the authority the lower the capital allocation. It is significant that, after correcting for population size, he found a strong correlation between the capital allocation and the size of the existing council stock, suggesting that the more rural authorities are discriminated against in the constraints on their investment. It will be seen that rural authorities with small stocks may be additionally disadvantaged in terms of their recurrent income.

Capital allocations are not a subsidy, but merely a limit on borrowing. However, a recurrent grant is paid by the Scottish Office to a minority of local authorities to help balance their HRAs. This subsidy is the housing support grant (HSG).

Local authorities have to raise money each year to pay the annual loan charges on previous building, and also to cover the expenses of repairs, maintenance and management. Loan charges are by far the largest item of expenditure. As shown in Figure 7.1, the revenue to meet these annual expenses comes overwhelmingly from council house rents, with perhaps a contribution from the rates and perhaps the HSG from the Scottish Office.

The housing support grant was introduced in 1977 as a form of deficit finance, with the intention of allowing authorities to raise their HRA expenditure without increasing council house rents: this was part of the then Labour government's attempt to control inflation. Thus, the HSG is calculated by the Scottish Office to fill the gap between the approved expenditure of the council and its assumed income. Since 1981, central government has been able to use this mechanism in reverse in order to force housing authorities to increase council house rents. In calculating the HSG, it has been assumed by the Scottish Office that rents are increased to the level desired by central government, thereby reducing the HSG subsidy payable. The scale of the reductions in the HSG is shown in Table 7.2.

Table 7.2 Housing support grant (constant 1985–6 prices, £m)

	Rural	*Remote*	*Scotland*
1979–80	49.75	34.04	342.44
1980–1	42.66	35.46	308.29
1981–2	25.81	24.29	198.65
1982–3	15.74	20.47	119.95
1983–4	12.03	18.24	79.16
1984–5	9.33	17.01	65.75
1985–6	6.92	14.34	48.24
1979–85	–86%	–58%	–86%

Source: Hansard, 5 December 1985, cols. 329–32

In 1979–80, the HSG constituted 39 per cent of HRA income on aggregate but by 1987–8 this contributed only 6 per cent, and by 1988 thirty one of the forty six local authorities in Scotland received no HSG whatever: while it is mainly the rural and remote authorities who still receive some subsidy to meet an expected shortfall of revenue, Table 7.2 demonstrates that they too have suffered a reduction. The rural authorities lost 86 per cent of their HSG between 1979 and 1986, and the remote authorities lost 58 per cent. A similar withdrawal of subsidy has taken place in England and Wales, where housing subsidy had fallen to only 7 per cent of HRA income by 1983–4, and five out of seven authorities received no subsidy at all in that year (Bucknall 1984).

Faced with this reduction, housing authorities had either to increase rents as central government wished, or to increase their subsidies from general rate funds (RFCs). The option of increasing RFCs was never likely to be popular with rural authorities, and in any case was precluded by strict financial penalties. If the RFC exceeded the Scottish Office's guideline, an amount equal to the excess was deducted from the HRA capital allocation; later legislation made it illegal to exceed the guideline (Rating and Valuation (Scotland) Amendment Act 1984). Further, high spending authorities might be faced with loss of rate support grant and ultimately rate-capping (now community charge capping).

To some extent the loss of income could be borne by reducing expenditure on repairs and maintenance, but rural councils had much less scope to do this than the urban authorities, for whom repairs and maintenance took up a greater proportion of spending. The combination of higher construction costs and more recently-built stocks in rural areas result in loan charges occupying a much higher proportion of rural authorities' HRA expenditure, thus reducing the scope for savings.

Increasingly, therefore, housing authorities have become more dependent upon rental income to finance their existing debts and any new building, and it is here that rural authorities are systematically disadvantaged. Because rural authorities generally have a smaller stock of council houses in relation to needs, the necessary rental income to replace the HSG must be raised from fewer tenants, so that if the HSG is withdrawn rents will have to rise more to finance rural housing programmes. This is quite apart from the higher cost involved in rural housebuilding, which has always required greater HRA expenditure per dwelling. Table 7.3 shows that council rents rose considerably faster in rural and remote areas during 1979–87. In 1979 rents in remote Scotland were 5 per cent above the Scottish average and by 1985 this differential had risen to 20 per cent. In rural areas of Scotland the average council rent had been 2 per cent below the Scottish average in 1979, but by 1985 it had risen to 13 per cent above the national average. However, urban rents have caught up a little since 1985. In rural and remote Scotland the rental base is so small, and becoming smaller through council house sales, that rental incomes are necessarily less in relation to the building programme required to meet needs: unless the HSG compensates for this shortfall, rural authorities must therefore find it progressively more difficult to meet their loan charges and to finance new building.

Table 7.3 Council house rents (cash prices)

	1979	*1981*	*1983*	*1985*	*1987*	*Increase (%)*
Remote Scotland	269	432	591	722	824	206
Rural Scotland	251	418	572	677	792	216
Scotland	256	400	513	601	762	198

Source: calculated from Scottish Housing Statistics

The upward pressure on rents is increased by, and contributes towards, the higher level of council house sales in rural and remote areas (see Table 3.1). As rents increase it becomes more attractive for tenants to buy their homes at a discount, thus reducing the number of tenants who are able to contribute towards the council's HRA expenditure: consequently additional rent increases follow which encourage further council house sales in a vicious circle. Ultimately, tenants will be paying high rents and the council may still have difficulty in balancing its HRA.

In the longer term, councils might relieve the upward pressure on rents by reducing housing investment and so allowing loan charges gradually to fall as debts are paid off. The rapid increase in rents has made councils wary of adding to their loan charges through major building programmes, even if their capital allocations permit. The outcome of both the reductions in capital allocations and the withdrawal of HSG subsidies has been that councils throughout Scotland are building markedly less houses: the scale of the decline is shown in Table 7.4 which enumerates housing starts by local authorities in each of the years since 1978–9.

It can be seen that overall council house starts now number less than 40 per cent of the total in 1978–9, itself a reduction from the high level of activity in the 1960s and early 1970s. This cutback in council building is likely to be intensified even further following the 1988 Housing Act's amendment of the 'cost-floor' rule. Prior to this, the discount on council house sales could not reduce the price to less than the outstanding debt still to be paid by the council on that property. Henceforth that protection lasts for only five years after investment in a new building or modernisation (eight years in England and Wales), and many authorities regard it as imprudent to make such an investment and risk a major financial loss after only five years. This fear will be increased by the advent of portable discounts.

Table 7.4 New dwellings started by local authorities

	Remote Scotland	*Rural Scotland*	*Scotland*
1978–9	599	1,300	4,195
1979–80	673	1,324	4,578
1980–1	548	1,004	2,694
1981–2	525	710	2,030
1982–3	652	942	2,372
1983–4	537	916	2,175
1984–5	398	493	1,705
1985–6	406	663	1,665
Decline 1978–86	–32%	–49%	–60%

Source: Hansard, 24 February 1987, cols 178–9

The consequence of these cutbacks in real housing investment since 1975 is that the new housing authorities have been unable to redress the housing problems which they inherited in 1975. Even in 1979, councils complained to the Scottish Office that they were unable to fulfil their statutory duties as a result. Several, like Argyll and Bute (1979, 11), commented in their housing plans that their capital allocations 'were considered inadequate to allow the council to carry out its agreed programme'. Stewartry District (1979, 10) went further:

> The preparation of housing plans has become nothing more than an academic exercise, in the course of which they [the Council] contemplate the policies and programme which they would have to adopt to fulfil their statutory function. This they do in the knowledge that such a programme has no prospect of being brought to fruition because it is unlikely that a sufficient capital allocation will be made available by central government in the foreseeable future.

Although this discussion has focused on Scottish rural areas, the general mechanisms operate similarly in England and Wales with identical results. In each case central government has imposed drastic reductions in housing investment on local authorities. Both north and south of the border, central government policies have sought to shift the cost of council house provision from the taxpayer on to the tenant, by forcing councils to increase rents in real terms: this change has necessarily discriminated against rural housing authorities, making it more difficult for them to provide council housing. Rural authorities incur greater

HRA expenditure per dwelling in order to meet existing debts and to build additional council houses. Despite this, a discussion of the alternative sources of HRA income has demonstrated that authorities are having to rely more and more on rental incomes, and that rural authorities with their small stocks will be least able to increase income from this source. Therefore, if rural and remote housing authorities suffer further reductions in subsidy, council housebuilding in rural areas is likely to become even rarer, quite apart from the reductions in capital allocations which constrain councils more directly, and the disincentive imposed through the alteration to the cost-floor rule. For this reason, it seems unlikely that housing authorities will be able to expand, or even maintain, their current level of council housebuilding in rural areas.

This conclusion is of crucial importance to both the principal themes of this book. A potential policy solution which had emerged to the conflict of objectives between landscape protection and housing provision required that local planning authorities refuse permission for private residential development where landscape quality dictates that only limited building is acceptable. Instead the authorities reserve the limited supply of available land for council housebuilding. Such a strategy is fatally undermined if local housing authorities are unable to proceed to build council houses because of centrally imposed financial constraints. This conclusion is also highly relevant to the theme of equity, since a decline in council housebuilding exacerbates the shortage of secure rented accommodation to which disadvantaged groups aspire.

The government, nevertheless, believes that there are options open to local housing authorities in rural areas. In a response to pressure from rural authorities for action to relieve second home problems in Wales, the Welsh Office retorted that 'local housing authorities already have open to them a variety of ways, in addition to themselves building new houses for resale, for assisting the provision of housing for local people' (Welsh Office 1982, 1). Unfortunately, from the point of view of local authorities, all the alternatives proposed by the Welsh Office involve the sale of housing to tenants rather than the provision of rented housing, unless the purchase of private dwellings for renting is financed from capital receipts from council house sales. Rural authorities tend to dislike such measures, questioning their relevance to the needs of poorer households, and instead have stressed the importance of preserving a 'reservoir of public housing' with which to serve local needs – an argument which seems to have been accepted by the Secretary of State in his recent announcements.

Despite the shortcomings of these measures, there are some alternatives open to local authorities which avoid the sale of council houses. These include housing association schemes (as noted already), self-build schemes and short-leasing schemes. Further possibilities may involve the private sector in providing rented housing, through a scheme of assured tenancies or some other form of social housing (Clark 1988).

Housing associations in rural areas

Housing associations were envisaged during the 1970s as the 'third arm' of housing provision, gradually replacing the declining private rented sector as an alternative to owner-occupied and council-owned housing (HMSO 1977). In the rural context, with the barriers of access to owner-occupation and the constraints on council provision already noted, and where private rented housing has in the past often formed a large proportion of the stock, housing associations may therefore have considerable potential relevance as an alternative form of social housing for disadvantaged groups. For this reason, the Development Commission in England has sought to encourage housing association activity in rural England linked to its own programme of employment creation, and more recently the DoE has supported the work of NACRT in seeking to establish rural housing associations in England and Wales. In 1988 the DoE announced a tripling of its support of NACRT's own administrative costs and a doubling of the Housing Corporation's funding of village housing schemes. Notwithstanding this encouragement, housing associations still provide only a very small fraction of the rural housing stock, although it has been argued (Bruce 1986) that in consequence of the diminishing role of local authorities, associations are likely to become the major providers of new rented housing in rural areas.

The finance for housing association schemes comes indirectly from the Scottish Office and the DoE, mainly (and increasingly) through the Housing Corporation. In the past a small part of the total also came from local authorities, but they are now less willing to divert resources from their own building programmes in the context of their severely reduced capital allocations of the 1980s. Thus local authorities have expected housing associations to rely almost solely on the Housing Corporation to provide them with loan finance. Further, in Scotland the Secretary of State implemented a revised system of lending to housing associations in 1985 which prevents local authorities financing them and makes the Housing Corporation (now Scottish Homes) the sole lending authority.

The Housing Corporation has suffered less from cutbacks in housing expenditure than local authorities. Thus, while it has a relatively small budget of £950m, this now amounts to 25 per cent of public sector housing investment (1987–8). Until the formation of Scottish Homes in 1989, the government determined the Housing Corporation's capital allocation each year, which was split between England, Wales and Scotland. The Scottish allocation was then divided between Strathclyde and the rest of Scotland (including most rural areas), hitherto on a 60:40 ratio. In the past, the corporation's first priority has been to assist the inner-city areas of Britain, and so spending has been heavily concentrated on these areas with only very small amounts allocated for rural schemes. However, there were signs latterly that the Housing Corporation was increasing its rural involvement, both in England and Scotland.

In its 1984–5 annual report the Housing Corporation (1985, 5) stated that:

> The Housing Corporation's aim is to direct funds to areas of greatest need. This means that around two-thirds of funding is allocated to associations working in the major conurbations and inner cities. However, we recognise that some rural areas suffer from problems which, although on a smaller scale, can be equally severe. This year in England we continued to jointly fund village housing schemes with the Development Commission and agreed to work closely with them in Rural Development Areas. We are also prepared to register a limited number of specialist county-based rural associations. In Wales about a quarter of the development programme was directed to rural housing. In Scotland there is a growing awareness of the housing needs of the smaller towns and villages, with some schemes supported in remote areas such as Skye.

However, these encouragements for rural housing schemes received a setback, according to NACRT, which pointed to a 1986 government directive to the Housing Corporation to allocate 87 per cent of its funds in future to inner-city schemes (NACRT 1987, 36). This threat appears to have been overtaken in the event by the government's enthusiasm for NACRT's approach of establishing village housing associations, which is discussed later.

In Scotland, the Housing Corporation reviewed its investment strategy in 1987, proposing to encourage locally-based housing associations to participate in a coordinated programme of rural regeneration (Young 1987). Previously the Corporation's limited rural programme had concentrated on providing housing for special needs (especially for the elderly), and this tended to be confined to the larger settlements. Since 1983, however, it seems that 'ministers have encouraged diversification of the programme into smaller settlements' (Young 1987, 86), and some evidence of this new attitude towards rural areas – (reflected in new building if not in rehabilitation schemes) is presented in Table 7.5.

Table 7.5 Housing Corporation funded schemes in rural Scotland

	Rural share of programme (%)		
	Rehabilitation	*New*	*Total*
Completed schemes 1974–86	3	23	10
Completed schemes 1974–88	3	25	12

Source: Young (1987); Watkins (1989)

This commitment was reaffirmed in the Housing Corporation (Scotland)'s 1987 investment strategy which identified rural housing as one of its priority areas for the next few years. A number of rural designated areas were being identified with the objective of stabilising population as part of a programme of rural regeneration. According to Young (1987), this objective embraced the corporation's desire to help stabilise rural communities both by helping existing residents to be well housed in their local area and by working in partnership with other agencies in a broader approach to rural regeneration. Notwithstanding the then minister's view that housing associations should concentrate only on special needs (SDD 1986), this strategy proposed that associations should, in addition, provide both rented accommodation for general needs and low cost home ownership. Some examples of the current breadth of housing association activity in rural Scotland are given in SFHA (1987).

To facilitate rural schemes the Housing Corporation (Scotland) revised its system of subsidies for site acquisition and development. Previously these were proportional to the total costs of the scheme, thus favouring larger schemes; but from 1985 a flat rate has been payable per scheme (£8,000) plus 1 per cent of allowable land costs and qualifying tender costs. As Bruce (1986, 50) explains,

> the effect of this change for rural areas is very beneficial, in that HAs are now being encouraged to 'think small' and can thus be seen as a positive step to meeting the needs of smaller rural communities where the need might be for a small scheme of 6/10 units.

Scottish Homes, the successor to the Housing Corporation (Scotland) from 1989, has been charged by the government with further extending housing associations' activities into rural areas (SDD 1987, 9) and this appears, indeed, to be the central element of the government's housing policy for rural Scotland. Housing associations thus seem set to become the major providers of new rented housing in rural Scotland, despite their small stock in the landward areas at present.

In England, too, the Housing Corporation has been receptive to requests for it to encourage more housing association activity in rural areas. In 1977 it agreed to finance up to 200 houses each year in areas where the Development Commission was building advance factories, provided the commission guaranteed the rents of any houses not occupied: however, only 130 units were built, rather than 600, in the first three years of the initiative. Partly this was because of an inability to assess housing needs in rural areas, partly because of the difficulties thought to be attached to implementing rural schemes, and partly because of the prior claims of urban areas hit by severe housing expenditure cutbacks (NACRT 1987, 36). Further, in many rural areas there simply did not exist the vehicles for attracting, or using, the available funding (NACRT 1987, 57).

More recently the Housing Corporation in England has attempted to address rural needs in two significant ways. Between 1982 and 1984 the Housing Corporation and the Development Commission jointly funded a programme of 100 shared ownership houses in sixteen village schemes. To meet the higher development costs associated with small, rural schemes the Development Commission made available an additional £5,000 per house, on top of the corporation's normal cost limits. While these schemes succeeded in making low cost housing available to local people who were unable to buy outright in their localities, the experiment has been regarded as unsatisfactory for several reasons (NACRT 1987, 61). In the first place, resales after the full equity has been sold were not restricted to local needs, and indeed one of the first houses has now become a second home. Second, the right to staircase (i.e. to increase the occupier's equity share) meant that the marketable equity shares became progressively larger, and hence further beyond the means of future first-time buyers. Third, high land costs excluded these schemes from the highest cost areas where they may be most needed. The suggestion of Shucksmith (1981, 132–5) that shared ownership without staircasing, giving the local authority an obligation to repurchase, would be most suited to the rural context appears relevant, and NACRT has also proposed limits on staircasing in any future extension of this scheme (NACRT 1987, 63). Despite many representations to this effect, and even a House of Commons amendment tabled by a former minister (June 1989), the government has refused to allow public money to be used in support of shared ownership schemes without staircasing. Therefore, NACRT uses only private funds for its shared ownership schemes, arranged through the English Villages Housing Association, specifically in order to prevent staircasing. This ensures that future purchasers have a similar opportunity to set foot on the ladder of home ownership at low cost. This experience is returned to on p.165, when considering private finance.

The other rural initiative taken by the Housing Corporation in England has involved supporting NACRT itself to coordinate a £5 million programme of building homes for rent, in small schemes of as few as four or five houses. In some cases existing housing associations have been used, but in many cases new county-based associations were established specifically to address the needs of villages.

> The new associations work only in villages and use local contacts and local knowledge to research the needs of the small communities. All are registered with charitable rules and all have committees drawn from local rural communities. They have no funds to employ staff and the work is shared between volunteers from the communities, local Parish Councils, staff of Rural Community Councils and the staff of NACRT. This combination of local voluntary effort and a pool of employed expertise has brought about a programme in the last two years of 22 new schemes (138 houses) which will be developed by the new Rural Housing Associations, with further schemes already in the pipeline. All the developments are for rent, 25 per cent for the elderly or the disabled and 75 per cent for young couples and families. Tenants will be selected from among those in housing need who are local to the village; there are already long waiting lists and it is obvious that all needs will not be met.
>
> (NACRT 1987, 37–8)

Despite this success, NACRT admits that

> This rosy picture of some 22 villages which will soon have some affordable housing for their own people should not obscure the difficulties that still remain. The new initiative is a small advance on the situation which existed two years ago and is a long way from being a widespread solution which can be applied quickly in many villages. There are large areas of the country where funding has not been made available, largely because of cuts suffered in the Housing Corporation's own spending, but also because of the [restricted] areas chosen for investment.
>
> (NACRT 1987, 38)

Indeed, only 1 per cent of the Housing Corporation's current investment programme in England goes into villages, and funding for just 305 new rented houses in fifty-three villages has been allocated in 1988–9. The target even for the new enhanced rural building programme announced by ministers in 1989 is only 1800 village homes per year in the early 1990s, to meet the needs of every village in England, and if this is to be the main thrust of the government's attempt to meet rural housing needs it appears quite inadequate.

These two initiatives in England, and recent policy developments in Scotland, would seem then to offer some hope of housing associations achieving their undoubted potential for helping disadvantaged groups in rural areas. However, as NACRT makes clear, the existing funding scheme in England continues to make it uneconomic for associations to build in villages, and their ability to carry loss-making projects is reduced by their overall shortage of funds. While an increase in activity from fourteen village schemes in 1986–7 to fifty three schemes in 1988–9 funded by the Housing Corporation is an improvement, it is clearly not enough to allow continuity of housing associations' programmes. Young (1987) also points to the problems arising from higher rural costs, difficulties of land acquisition and assembly, the inappropriate nature of urban-derived policies for assisting rehabilitation, and the additional burdens and costs involved in attempts to coordinate housing schemes with other agencies in pursuit of integrated rural development strategies. Each of these difficulties emphasises the need for further experimentation and additional resources, and these are subject to central government control.

A further uncertainty surrounds the future arrangements for the finance of housing association schemes. Whilst housing associations are self-governing, as already noted they are funded almost entirely by government. The government has made it clear that housing associations are expected in the future to rely increasingly on private funds. In the past, over 90 per cent of the capital cost of rural schemes has been paid directly by the Housing Corporation in the form of Housing Association Grant (HAG), so allowing rents to remain at fair rent levels. In future, housing association tenancies will be assured tenancies with rents rising towards market levels, and a greater share of the investment capital is likely to have to be raised from the private sector. A pilot project operating in Scotland in 1988–9, for example, was subject to a maximum level HAG of 75 per cent, and no rural schemes were viable under these conditions. Even at higher levels of HAG it is likely that the rent levels necessary to repay private investors may rise beyond the means of present client groups, so calling into question the housing associations' *raison d'être*. If rural associations are given sufficient HAG to enable them to charge rents affordable by their present client groups, then any expansion of rural activity will require increased public investment (Shucksmith and Watkins 1990).

> The future of Housing Association activity in villages lies largely in the hands of Government. The Rural Housing Associations and some other Associations have demonstrated that, with the encouragement of funding being offered, they can identify and progress new schemes quickly. The funding for rural schemes is still barely

a trickle ; unless it increases to a steadier flow, there is little reason to be hopeful about the future.

(NACRT 1987, 39)

Once again it seems that central government exercises crucial control over housing associations' potential for resolving rural housing problems, as it did over the actions of local planning authorities and over local authorities' programmes of council building. Whilst housing associations are independent and self-governing, they are funded almost entirely by government and are subject to the detailed control of a central agency, the Housing Corporation or Scottish Homes, which themselves are subject to the direction of central government.

Furthermore, there is a contradiction inherent in the government's eagerness to expand the contribution of housing associations to the extent that they become the major providers of social housing, while at the same time wishing to increase the contribution of private investment and contain public expenditure. The housing associations' role in providing low cost housing for disadvantaged groups in rural areas can only be expanded through a substantial increase in public investment.

Private finance

In an attempt to escape from this strait-jacket, some housing associations have now begun to examine ways of attracting private finance to allow them to expand their provision of rented accommodation. In 1982, the National Federation of Housing Associations in England suggested that it might coordinate private sector finance, underwritten to the sum of £500,000 by the Development Commission, for locally promoted housing association schemes in rural areas (NFHA 1982). Private investors would finance the building work by buying a share of the capital value of the completed houses, so enjoying an appreciating capital asset and an increasing income return. Unfortunately, no such scheme appears to have been realised, and, in any case, it seems that such funding would be regarded by the Treasury as public spending if a public agency bore the risk.

More recently, the government's White Paper, *Housing – The Government's Proposals*, signalled an intention gradually to replace council housing with 'social housing', financed, provided and managed by a multiplicity of alternative landlords, including housing associations, cooperatives, building societies and trusts, financed by pension funds and other private sources. The Housing Act 1988 has specified that all new tenancies are to be at open market rents, under assured tenancies or short assured tenancies, and the government hopes that this rent decontrol will encourage a revival of the private rented sector.

Further encouragement is given through the Business Expansion Scheme (BES), which provides investors in rented housing with generous relief from income and capital gains tax for five years. According to the Secretary of State (DoE 1988a),

> taken together these measures will transform the prospects for investment in housing for rent. In rural areas in particular they may have a significant impact, especially where there are landowners or farmers with redundant buildings or spare land which they would prefer to devote to the benefit of their local communities if only they could do so and make a reasonable return.

Yet it seems unlikely that many private investors will be attracted to build rented housing for lower-income groups in rural and remote areas: construction costs are high and rent levels might be expected to be low, given the greater incidence of low incomes. A worthwhile rental income would only accrue to the landlord if levels of housing benefit were so high as to constitute a 'permanent and inflation-linked subsidy to successive generations of tenants' (NACRT 1987, 30), and it seems improbable that rent officers would accept the rents required by investors as reasonable.

NACRT argues that such provision would be inefficient compared to investing directly in new houses. Whitehead and Kleinman (1986) also believe that additional rented housing would be most efficiently provided by local authorities and housing associations, and that

> the alternative of inducing supply by private landlords via large-scale payments through the housing benefit system can only be justified if there is reason to believe that the private sector can better meet the requirements of households in that tenure.'
>
> (Whitehead and Kleinman 1986, 172)

This, they argue, is not the case. The investor is therefore unlikely to be able to earn a commercial return on his capital from building houses to let at affordable rents in rural areas, except perhaps in the main centres. The English housing minister, the Earl of Caithness, appears to be pinning his hopes for affordable rural housing on the tax incentives offered through the BES, which he viewed in October 1987 as a 'powerful incentive' for investment in rural areas. However, none of the prospectuses issued so far appears to promise any low cost housing in rural areas, and instead the emphasis is firmly on the prosperous young, the mobile and the elderly at the upper end of the urban market. Constable (1988) believes that in any case 'rents will be well beyond those on rural incomes' and that the BES will not even scratch the surface of rural housing problems.

Nevertheless, NACRT believes that assured tenancies may have a role in rural areas, but as a complement to fair rent housing rather than as its substitute. Assured tenancies would be appropriate to the circumstances of tenants whose means lie between highly subsidised fair rent housing and owner occupation:

> such groups are often the most disadvantaged in housing terms; they can just afford private rents in accommodation that is just tolerable, but it is often temporary or overcrowded. Their accommodation is deemed sufficient to give them a low priority on housing waiting lists and they see no way of improving their housing conditions.
>
> (NACRT 1987, 31)

Given adequate safeguards, assured tenancies additional to the essential public sector programme are cautiously welcomed for these groups by NACRT (1987,31). The only fundamental objection to such an application of the assured tenancy would seem to be the concomitant residualisation of the public sector, and an increasing tenurial stratification of the rural population.

Some housing associations take a contrary view of private finance, however, and are happy to rely on housing benefit to pay the resulting higher rents for their traditional client groups. Since a very large majority of housing association tenants are eligible for full housing benefit, it is argued that rent increases are painless and of no concern: it follows that such associations will be able to attract private finance for assured tenancy schemes in rural areas for the poorest groups in society. Those who have explored this option appear to have found no difficulty in finding willing lenders among the banks and building societies.

For all but the poorest, a more feasible means of attracting private finance into the provision of low cost rural housing is through shared ownership schemes without staircasing, as mentioned previously. Such schemes are already undertaken by the English Villages Housing Association, generally using 100 per cent private finance. The key to this approach is again through the acquisition of low cost land, although other means of reducing the cost could be found.

If land costs are reduced through the planning mechanisms outlined on pp.145–6 such that a house can be built for, say, 80 per cent of its open market price, then a housing association has considerable scope for retaining a share of the equity and so retaining a degree of control over the house. Instead of selling the house outright, the housing association may sell only 80 per cent of the equity to the occupier and retain the remaining 20 per cent themselves. The occupier will have paid only 80 per cent of the open market price, and when he sells he will receive 80 per cent of the then market price: he will thus occupy the house at low cost, while still benefiting from any increase in the house's value.

Meanwhile, the housing association will have borrowed short-term private funds to build the house (80 per cent of its value) which are then repaid in full once 80 per cent of the equity is sold to the first occupier. The remaining 20 per cent is held by the housing association, allowing it to retain control of the house and thus to ensure that subsidised access to the house will be available to subsequent purchasers, and not only the first one.

Apart from avoiding staircasing, and so ensuring that the housing remains affordable in the long-term, the scheme has the advantage that homes may be allocated by the association according to housing need. Most importantly, this is achieved with private funds additional to the insufficient public investment: indeed it appears to be quite easy to obtain private funding for such schemes, since they are in effect secure short-term loans. Unfortunately, however, this is only currently possible in areas of the country where land costs form a substantial proportion of the total cost of housebuilding and where cheap land can be obtained. Similar schemes would not be feasible in areas of northern Scotland, for example, where the land element is a small proportion of the total cost, or where the costs of construction may even exceed the market price of the completed house. To make shared ownership without staircasing feasible in such places another source of subsidy would be required.

Other initiatives

Short-leasing schemes

The housing minister also foresaw local authorities shedding their role as direct providers of housing, adopting instead an 'enabling role', acting as a focus for the matching of a variety of problems and solutions. In one way, rural authorities have already shown themselves capable of operating in this mode, through the widespread short-leasing schemes designed to bring together empty private houses and tenants from waiting lists.

These schemes were pioneered in north Wiltshire from 1974, and were soon adopted by many other rural authorities in England and Wales. In Scotland the approach required a change of legislation in 1980, and this was followed by a Shelter project, financed by the Scottish Office, to promote the adoption of such schemes in Highland and Grampian Regions (Shelter 1985). In either case, the scheme is the same: in many rural areas there are many empty, but lettable, houses whose landlords were discouraged from letting them by the security of tenure provisions of the Rent Acts and perhaps also by the expense of improving and managing the properties. However, under Section 5 of the Rent Act 1968 (in England), and under paragraph 14 of schedule 2 to part 1 of the Tenants' Rights (Scotland) Act 1980, a sub-tenancy is unprotected where the immediate landlord is the district council. Thus, if a private landlord lets a house to the council, and then the council sub-lets it to a tenant from its waiting list, the landlord may recover possession from the council on the expiry of the lease.

District councils have modified the scheme in various ways, but generally it is applied as follows. The owner lets the property to the council at an equivalent council house rent for an agreed term, usually of at least a year, with provision for the lease to be terminated any time thereafter, given a stipulated period of notice. Where the house requires improvement work involving receipt of an improvement grant from the council, it would be usual for the council to negotiate an agreed term of at least five years. The council then sub-lets the house on its own terms to a tenant from the waiting list on the basis that the tenancy is short-term, and guarantees to provide alternative council accommodation for them when the lease ends. The council also undertakes the management and maintenance of the property and guarantees to return it to the owner in good condition, with vacant possession, at the end of the lease. Thus the owner receives an income from an otherwise empty house, perhaps has it improved, avoids the burden of management, and is certain of recovering possession when he wishes. Meanwhile, a tenant is housed who would otherwise have been homeless.

The scheme has several attractions. In the first place, it is highly cost-effective, costing the council virtually nothing to house additional applicants from their waiting list. Second, it conserves buildings in rural areas, arresting the deterioration of vacant houses and in many cases rehabilitating unfit houses. Third, the scheme involves no new building, and thus helps to meet housing needs without harming landscape. Most importantly in the present context, because it involves virtually no capital or current spending, unless improvement to the house is necessary, the scheme allows councils to increase their effective rented stock beyond the control of central government.

The scheme has been a clear success in many rural areas, but there are two limitations on its operation: it is not likely to be a success in every area, and furthermore it is necessarily limited in its size. The success of the scheme depends upon the existence of vacant properties which might be let were it not for the security of tenure provisions of the Rent Acts. In many rural areas, however, there is already a form of let which avoids all worries about repossession, and this is the holiday let. In attractive rural areas, and especially in national parks, AONBs and coastal areas, the income which can be earned from seasonal holiday lets greatly exceeds the potential return from council rents. Thus, in Eden

District a similar short-let scheme was markedly unsuccessful: only one house, in Penrith, was let during the three years of the scheme. The short-leasing scheme is of little relevance to such areas, where housing problems may be most acute because of the additional pressure on the housing stock.

The other constraint on the scheme's operation derives from the necessity of the council being able to guarantee to rehouse the tenant when the lease expires. As a report from North Wiltshire District Council (1977, 4) indicates:

> It is crucial to the success of the scheme that the Council should not put themselves in a position whereby they cannot honour their obligation to the owner to rehouse the sub-tenant at the expiration of the tenancy...... [The number of houses in the scheme, therefore,] will largely depend on the extent to which the Council will be willing and able to build and convert, and the extent to which the turnover of existing accommodation meets demand, rather than the availability of suitable properties for leasing.

The scheme's size therefore depends on the rate of new building and the availability of relets, to ensure that the council's guarantee can be met. At present, two elements of central government policy combine to undermine the increased use of short-leasing schemes. Financial constraints prevent local authorities from embarking on new building of council houses; and mandatory council house sales have reduced the turnover, and the pool, of existing council accommodation. Both these policies reduce a council's ability to rehouse tenants at the expiry of a short lease, and councils must therefore be wary of increasing, or even maintaining, the number of properties within its scheme.

Nevertheless, Alexander (1987, 96) argues that even if only a small number of houses can be brought into these schemes, 'in a period which has seen several years of severe financial restraint, a leasing scheme may well be the most significant way a local authority can provide homes for those on waiting lists', and he instances the position of Lochaber District Council who are fairly typical in being unable to build any new council houses, but now rely exclusively on a leasing scheme to increase provision. However, while the numbers of houses are important at a local level, the aggregate number is small and necessarily limited. It is apparent that even short-leasing schemes depend upon continued central government investment in council housebuilding, and that in current circumstances the contribution which such schemes can make is small in relation to aggregate need. A further constraint applies where the houses offered require substantial restoration, since such spending also derives from the council's capital allocation.

Municipalisation, homesteading and nominated leasing

Apart from their cost-effectiveness, a major attraction of short-leasing schemes is their use of existing houses to increase provision without any damage to the landscape, thus offering a reconciliation of the conflicting objectives of rural policy which have been a central theme of this book. A similar reconciliation would also be achieved by a number of other means of diverting the existing stock to meeting the needs of disadvantaged households, such as municipalisation, homesteading and nominated leasing.

Municipalisation involves the purchase of existing private houses by the local authority for letting to households on its waiting list. Skye and Lochalsh is one authority which has pursued this approach, buying both an entire housing estate where this was (surprisingly) priced below the cost of council house construction and also 'secondhand' houses which compare very favourably with the cost of new building (Noble 1986). However, the experience in the Lake District differs: while this would be a particularly appropriate policy option in a national park, Allerdale, Copeland and South Lakeland have found municipalisation a very expensive means of providing housing to rent. This is compounded by the present structure of central government subsidies which gives the local authority a smaller exchequer contribution towards the purchase of a private house than for new building (Shucksmith 1983).

Homesteading has been pioneered in problem council estates, by the GLC (while under Conservative control) and by Glasgow District Council (Labour), to deal with difficult-to-let housing. These houses have been offered for sale to private buyers, usually from the waiting list, at very low prices: the buyers then rehabilitate their homes with the help of improvement grants. Noble (1986) has proposed that in the context of Skye and Lochalsh the concept could be applied to substandard private houses: the council would purchase such properties, by compulsion if necessary, and offer them for sale to waiting list applicants on a similar homesteading basis. People on the waiting list might even be encouraged to identify suitable vacant properties in their area which they would be prepared to take on as homesteaders. It is not clear whether such schemes would require additional capital allocations to be made available to local authorities, or whether expenditure would be balanced by receipts, but finance would appear to be the major obstacle to the development of this approach. Again, encouragement might be given to experimental schemes.

More radically, in many of the areas where the conflict between landscape protection and housing provision is most acute, a large proportion of the housing stock is euphemistically known as 'non-effective', in that it is occupied for recreational purposes rather than for meeting housing needs. A highly efficient, but probably unacceptable, solution to meeting housing needs in these areas would be through nominated leasing of second and holiday homes in suitable locations to households on the waiting list. While not quite so drastic as compulsory purchase, a solution to housing needs and landscape protection based on the imposition of tenants on private owners would involve a major redistribution of property rights, to say nothing of the effect on local economies of the lost income from tourism.

Self-build

One curious omission from most discussions of rural housing provision is that of self-build, where the occupier builds his own house. While this form of provision is the norm in the crofting areas and in many other countries, it is almost unknown elsewhere in Britain. Presumably the main reason for this is the difficulty of gaining access to land on which to build, and of obtaining planning permission, although no research has been undertaken to establish the reasons. Indeed the general ignorance of self-build schemes is apparent even from one of their advocates, who states incorrectly that no other authority in Britain has emulated Lewisham's self-build housing association (Ward 1987).

Some experiments in community self-build have been reported (Kinghan 1981; Clark 1981), and these have usually involved the formation of a housing association by a group of local people who intend to build houses for each member of the group, using their own labour. As Clark (1981) notes, the advantage of such schemes is that the necessary skills may well be present amongst manual workers in rural areas, allowing houses to be constructed at low cost because of the saving on labour: it is also particularly suited to the use of small building sites and to meeting dispersed needs. However, its success depends upon the skills and enthusiasm of the group, and upon the support given to them by, for example, the local authority, who may assist such societies with loans, serviced sites, training, lending equipment and bulk-purchasing materials.

An example of such a scheme is in Hampshire, where Hart District Council set aside land in 1977 for a self-build housing association to purchase and to build upon (Clark 1981). The council advertised in the local press and circulated those on the housing waiting list, and then selected among the applicants to ensure a range of building and other skills within the group. (The NFHA advises that about half the group should be skilled in building trades.) Once selected, the group was helped to form itself into an association. Thereafter, while the council maintained a close liaison, the project was run as any other housing association by the group members. Although the land had to be purchased at its full market price, the fourteen houses were built at low cost because of the saving on labour, and once all the houses were completed each individual occupier was able to purchase his house from the association at cost (BBC 1979).

The major difficulty to be overcome by such schemes is to obtain short-term finance to buy the land, materials and to pay for specialist advice. Two potential sources of finance in better times would be the Housing Corporation and the local authority, but as we have seen these are unlikely to be able to offer any significant funding. Building societies tend not to lend money until their loans can be secured on the finished houses, although this may change following the granting of wider powers to building societies by the Building Societies Act 1986. Even with finance available, Clark (1981, 4) has concluded that:

> Self-build is not easy. It places great strain on the self-builder's family life and first job... And self-build cannot meet the needs of many older, or lower-income families. It is not a substitute for fair rent housing from local authorities or housing associations.

These difficulties and frustrations apply also to individual self-build. In Britain, only a minority of houses in the countryside are built by their occupiers, and virtually all of these would be on infill sites within villages. Occasionally a local authority might sell serviced plots to individual self-builders as an experimental means of encouraging home ownership, but such schemes are not widespread. In Berwickshire, one such scheme was implemented at Gillsland, Eyemouth, but far from aiming to help lower-income groups the council auctioned each plot to the highest bidder, achieving land prices above the district valuer's valuation in each case, and resulting in expensive and socially exclusive houses. Argyll and Bute District Council, in contrast, has a pilot project at Port Charlotte, Isle of Islay, where it is intended that serviced sites will be sold at the district valuer's valuation to council tenants and individuals from the waiting list, subject to each house built being used only as a principal residence (Planning Exchange 1984). Several other local authorities, such as Badenoch and Strathspey, and Skye and Lochalsh, have also provided serviced sites where individual owners may build, but the success of such schemes is unresearched.

What is clear is that self-build schemes, whether individually or collectively organised, make only a tiny contribution to total rural housebuilding in Britain. This is in strong contrast to the experience of other rural areas in Europe, let alone the Third World where self-build housing is increasingly seen as the solution to the housing problems of the poor (Bryant 1980). In Ireland, most new rural house construction is of single houses built in dispersed locations on land obtained from a relative, and about 55 per cent of rural housing is entirely or partly self-built (Jennings 1987), despite planning controls designed to prevent this. This facilitates the housing of lower income groups with some local kinship tie who are able to reduce construction costs not only by providing their own labour but also by obtaining the site at no cost in many cases. In Norway and Sweden the common practice is also for new households in rural areas to build on land belonging to relatives and neighbours (Folkesdotter 1987): indeed, in Sweden a special fund was created by the state from 1904–57 to assist 'less well-off working people' to buy land to build on, and new single house construction remains heavily subsidised. Again, the effect is to encourage people who can obtain a plot of land to build their own home; and since they are most likely to be able to obtain land, people with relatives or other contacts in the countryside have the best chance of building such a house (Folkesdotter 1987).

Two factors determine the availability of sites for single houses in the British countryside. One is clearly the ownership of land, which is far more concentrated in Britain than elsewhere in Europe. Partly because of continuing farm amalgamations which have reduced the number of small farmers, partly because of the persistence of landlord–tenant structures and partly because of Britain's much earlier agricultural revolution and urbanisation, few families in Britain today still maintain kinship ties with rural landholders. The possibility of gaining access to land on which to build is therefore a matter of market strength rather than of kinship. Second, even if one has obtained a small plot of land, one has still to obtain planning permission, and as already noted post-war planning policy has successfully sought to prevent sporadic housebuilding in the British countryside. For both these reasons, apart from farmers and crofters, access to land for single housebuilding in rural areas has been largely confined to those with the wealth and income to compete successfully in the market for the few sites acceptable to planners, in contrast to the continental experience.

Any rural housing strategy which sought to promote self-build by lower income groups on a larger scale would, therefore, have to address both land ownership and established planning policies. Local authorities might use their powers to purchase land, by agreement or compulsion, to make a wide variety of dispersed, serviced plots with outline planning permission available for sale at valuation to households from their waiting lists. Collaboration with regional or county councils in servicing the plots would be essential, as would close supervision and support (Noble 1986). This would require additional staff input. Further, the ability of lower-income groups to participate in the scheme would probably require central government to offer cheap credit to self-builders, perhaps along the broad lines of the Crofter Housing Grants and Loans Scheme, which has been shown to be particularly cost-effective in public expenditure terms (Shucksmith 1990). Despite this, ministers have argued against extending this scheme to non-crofters because of the need for restraint in public expenditure (Noble 1986; Mackay 1987).

Conclusion

Constraints on public expenditure and other central government controls obstruct virtually all the avenues open to local authorities to increase the housing available to disadvantaged groups in rural areas. Conversely, several avenues have been seen to offer possibilities of better housing opportunities if investment in rural housing were allowed to increase.

A discussion of the various alternative policy responses to the apparent conflict of policy objectives between landscape protection and housing provision, and to the lack of housing opportunities for disadvantaged groups in rural society, suggests several suitable approaches, each giving different patterns of advantage and disadvantage, and each with different implications for landscape protection. Where landscape quality dictates that only limited new building is acceptable, the preferred solution would require the local planning authority to refuse permission for most private residential development, instead reserving the limited supply of available land for social housing for rent. This might be combined with increasing municipalisation of the existing stock, according to the pressure of need, and with short-leasing schemes, where feasible. However, such a strategy has been seen to be fatally undermined, either if planning authorities are not permitted by the Secretary of State to discriminate between private housing and social housing, or if housing authorities and housing associations are unable to proceed to build social housing because of centrally imposed financial constraints.

Where landscape quality is less critical and new houses may be built now that agricultural imperatives are reduced, a wider spectrum of alternative strategies might be applied. Again, the provision of social housing is likely to be a central component of any strategy because of its efficiency (Whitehead and Kleinman 1986) and its much greater equity. But this might be combined with widespread self-build initiatives, supported by local authorities ensuring the availability of suitable serviced sites, and encouraged by cheap credit from central government, targeted at those on waiting lists. Private speculative developments would be unlikely to meet the needs of disadvantaged groups, but in many areas of countryside, including most of the Highlands and many other undesignated areas, there might be no planning objection to such developments proceeding alongside social housing developments and self-build schemes. In the remoter parts, where empty, sub-standard houses remain, homesteading experiments might be tried together with a more active promotion of improvement, as outlined in Shucksmith (1990) and Shelter/Rural Forum (1988). In some areas there may be merit in assured tenancies, subject to the conditions listed by NACRT (1987), but these are unlikely to make a significant contribution for the reasons stated previously.

However, these policy options are currently precluded by the imposition of central government constraints on investment: most clearly, this applies to the provision of council housing, but it has been demonstrated in this chapter that it also extends to provision by housing associations, short-leasing schemes, self-build, municipalisation and, perhaps, homesteading. It even applies to assured tenancies, insofar as these will require a long-term subsidy to occupiers to enable them to pay open market rents. Therefore, while the most appropriate form of policy response may vary from one rural area to another, a necessary precondition of that response will almost certainly be a relaxation of the constraints imposed by central government on public-sector housing investment.

Chapter eight

The central state: current concerns and future scenarios

Introduction

In Chapter 1, the Secretary of State was quoted as seeing himself on the horns of a dilemma, due to widespread concern about the need for low cost rural housing and also the effects of further building in the countryside. This chapter reviews the political dilemma facing the government as it seeks to devise a planning policy for land release compatible with its housing policy reforms. This involves a brief account of the debate over land availability, an attempt to unravel the corporatist liaisons of the central state and a discussion of the relationship between the central and local states. These, of course, are major topics deserving of a much lengthier treatment than is possible here (Shucksmith and Watkins 1988a–d).

The future is then considered through the elaboration of alternative planning scenarios. Three possible scenarios are assessed in terms of their capacity for reconciling the contradictions of policy objectives already identified, and in terms of their distributional effects. These are:

1. a general relaxation of planning controls to allow more house-building in the countryside;
2. the development of new settlements while constraining the growth of existing towns; and
3. releasing land only for low cost social housing.

While there is scope for a variety of forms of development to proceed, on the basis of circulars from the DoE and the statement on *Housing in Rural Areas: Village Housing and New Villages*, the three considered here are seen as analytically distinct.

Land availability

The debate on housing land availability has become one of the most controversial issues in recent years, as housebuilders and planners have disagreed over the amount of land needed to satisfy housing demand and over the balance to be struck between housebuilding and environmental protection. The debate has recently been fuelled by the release of household projections by the DoE, based on updated 1985 figures, which indicate that there will be an increase of 11 per cent (2,039,000) in households between 1986 and 2001. The largest increase in households (21 per cent) will be in East Anglia (particularly in Cambridgeshire at 30 per cent), followed by the Southwest at 17 per cent, the East Midlands at 14 per cent (Northamptonshire at 22 per cent), and the Southeast at 14 per cent. Figure 8.1 clearly illustrates this 'ripple' effect of growth in households beyond the Southeast, and indicates the increase in housing demand which the rural and shire counties will experience (DoE 1987).

Initially the debate on the availability of housing land centred on two main arguments: that the total area of land available for development in plans is insufficient and that inadequate 'brownfield' land is available in urban areas to make any impact on the scale of new land allocation requirements. Although several reports (JLRC 1984; SCLSERP/HBF 1984) have indicated that an absolute shortage of land for residential development does not exist and that structure plan allocations are sufficient, the recently estimated growth in households causes some concern as most structure plan provisions were formulated before these demographic changes were anticipated (King 1987). The latest DoE progress report on land for housing reveals that, for those counties where a comparison between their proposals and approved plans can be made, only five counties in England are planning to increase provision in their structure plan alterations, and '18 counties propose to reduce provision by an average of 16 per cent: 8 of these were in the south east' (Hooper, Finch and Rogers 1988).

The need, or not, to allocate more land for housing is clearly a more complex issue than calculating the total amount of land available for projected housing need. 'The distribution (or marketability factor), ownership and immediate availability of land for housing is more important than the total amount of land available for housing' (Herington 1984, 136). A DoE/HBF survey in Greater Manchester showed that much of the land for which planning permission was obtained was in the wrong position or of the wrong type to meet housing demand: out of 27,500 plots the survey suggested that only 17,000 were suitable for development in practice (DoE 1978). A more recent study by Thomas (1987) found that 'there is much less inner city land which is suitable and genuinely available for house- building than is widely assumed'.

Figure 8.1 Projected growth in households to 2001

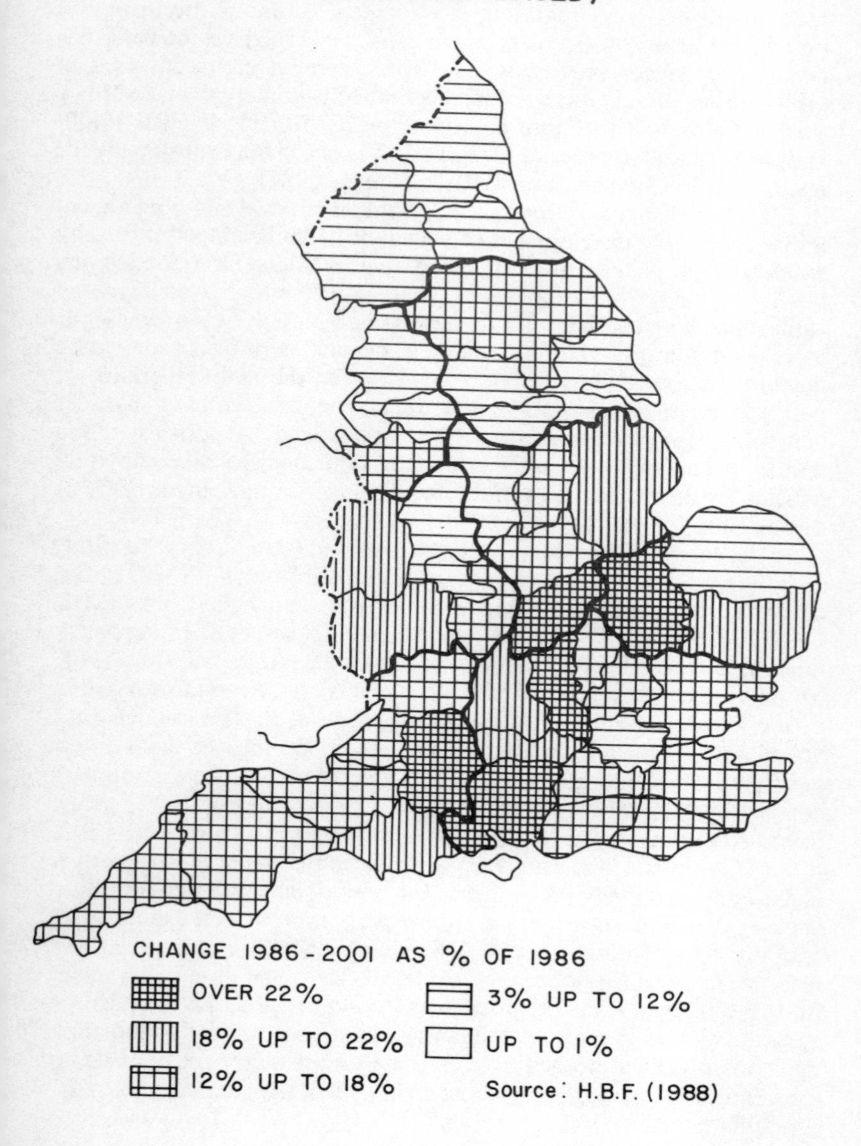

So although DoE household projections have been criticised as being unreliable by the CPRE (*Guardian* 28 May 1988), more land will clearly be needed for housing development to satisfy demand in the 1990s. Research on recent land use change in England (Roger Tym 1987) reveals that much new housing development is already occurring on brownfield sites (46 per cent of all land developed for housing has occurred on either previously developed, derelict or vacant sites in urban areas), yet this source of land is limited and likely to make only a small contribution to future housebuilding (JLRC 1984; HBF 1988). Despite the large reserves of land in urban areas, much is unsuitable for residential development (JLRC 1983; Thomas 1987).

It is clear that despite the success of urban renewal and government initiatives for the development of vacant or underdeveloped inner-city areas, the projected future housing requirements far outreach the possible supply of brownfield sites. There will therefore be an increased call to allocate greenfield sites for new housebuilding. At the same time, increasingly less agricultural land is needed to produce the same quantities of agricultural produce, and indeed the new imperative of policy is to reduce food production. This coincidence of circumstances is seen by many as an opportunity to reassess land use policies, and to permit development to satisfy housing demands, to alleviate rural housing problems and to relieve the budgetary crisis in the EC, as discussed in Chapter 1.

In October 1987, government circulars (DoE 16/1987; SDD 18/1987), followed by planning policy guidance notes (PPG2, PPG3, PPG7) in England and Wales and national planning guidelines (SDD 1987) in Scotland, all indicated a change in emphasis of planning policy relating to residential development on agricultural land. Buckley (1987, 3) summarises the release of the circular as 'a significant downgrading of the general importance of protecting farmland'. The circular provoked both a dramatic and a varied response. On the one hand critics have predicted that a great rush of development will occur and much countryside will disappear under large scale housing schemes, whilst others have anticipated benefits through decreased house prices as a result of increased land supply. Again, this has to be put in perspective, and it was noted in Chapter 1 that the amount of land required for future housing development is only a minute proportion of that estimated as 'surplus' to agricultural needs. Estimates of the additional greenfield land needed to fulfil housing requirements up to the year 2001, range from 81,000 ha. to 250,000 ha., compared to the notional 'surplus' of agricultural land of somewhere between 2.6 million and 5.5 million ha.: the amount of land needed to satisfy the estimates for future housing requirements is, at most, only about 2 per cent of the predicted 'surplus' agricultural land.

The housing land availability debate refers to regional and national housing requirements, of course, rather than to the lack of low cost housing (for rent and sale) for rural people. Nevertheless, the two are interrelated. The Secretary of State for the Environment, for example, has expressed the hope that private developers 'will pay increasing attention to the needs of local communities in rural areas' and build 'balanced developments so as to ensure that they provide for community needs for low cost housing for rent and for sale' (DoE 1988a). The discussion paper accompanying his statement made clear that proposed developments would be more acceptable 'if they offer a reasonable choice of housing types, costs and tenures which will help to meet local housing needs as well as those likely to have a wider market'. They are also interrelated in the housing market, insofar as limits on supply have the effect of increasing house prices and land prices to the exclusion of lower income groups, as discussed at length in Chapter 3. However, the mere comparison of projected household growth with land zoned for housebuilding takes no account of market mechanisms and of different social groups' means of access to housing. The land availability debate has not been concerned with the allocation of housing.

It is now generally accepted that more land is needed. The debate is over where housing development should occur and which land should be allocated for new housebuilding, the choices essentially revolving around more greenfield sites or further infill and suburban development, inner city or outer city, north or south-eastern England. The issue may have appeared in 1988–9 to be merely a battle between the past and (then) present Environment Secretaries, Michael Heseltine and Nicholas Ridley, but the issues raised are far wider than this. The conflicts and tensions between conservation and development, inner-city decay and outer-city/suburban expansion, southeast regional growth and prosperity and northern England's decline are all aspects of this debate, reflected in the involvement of many interest groups representing land investors, farmers, housebuilders, conservationists and planners (Ambrose 1986).

The role of the central state

Since the 1947 Town and Country Planning Act, land use planning policy has focused on the objectives of landscape protection and the protection of agricultural land at the expense of the objective of rural housing provision. Since responsibility for rural policy falls mainly into the remit of one ministry – the DoE in England and the SDD in Scotland – the reconciliation of the competing objectives in rural policy is the concern of the Secretary of State for the Environment and his Scottish equivalent, the Minister for Home Affairs and the Environment.

In a reply to a letter sent by the former Environment Secretary, Michael Heseltine, criticising the large-scale development occurring in the southeast, the then Secretary, Nicholas Ridley stated,

> I am glad you acknowledge that my unenviable task of balancing the claims of economic growth with environmental protection is a difficult one....It is the function of a land use planning system to meet the needs of economic development while protecting the countryside....We are strongly committed to our Green Belt and Countryside policies but we have to face the fact that the rate of population growth and household formation in the South-east is increasing and must be catered for....It is the responsibility of local planning authorities to ensure that necessary development is accommodated in ways that do not conflict with our policies for the Green Belt and the protection of the countryside and yet do not place unnecessary constraints on economic activity or on the desire of people for a home of their own....It remains the case, however, that it is simply not practicable to accommodate all development needs by building in the inner cities or by forcing people and firms to move out of the South-east to other parts of the country. Some new land has to be allocated for development in the South-east and I for one am not prepared to rule out all such development which is necessary to cater for legitimate housing needs. If I did, I would then be blamed – with some justification – for the consequent lack of reasonably priced housing for sons and daughters of local people, and for rising levels of homelessness. It is easy to point to 'greedy' developers. But they only exist because they have customers.
>
> (*The Planner*, Supplement, April 1988, 3)

The competing objectives and contradictory concerns of the central state are readily apparent in such a statement. But how is policy formulated by the central state in this context? In particular, what is the effect on central state policies of the many interests organised at national level, some of whom will be in favour and some opposed to new building?

At a central state level, a corporatist conceptualisation of policy formulation appears most fruitful in the context of planning and housing. Policy may be seen as the outcome of bargaining between the state and the larger business and commercial interests, which are incorporated by government into the policy formulation process in order to secure their cooperation in implementing policy. In terms of the policies relevant to land release, farming interests have long been incorporated into the policy process, following the need for farmers' cooperation in providing food for Britain. The NFU, however, has now had its corporatist influence diminished as the government commitment to expanding agriculture has been lessened. Lowe (1988, 39) suggests that the NFU's corporatist links are now only maintained by the government's need for their cooperation in intervening to reduce agricultural surpluses. As the corporatist power of the NFU has receded, the developers (represented by the HBF), and especially the volume builders, have come to prominence, largely because the cooperation of private housebuilders is essential for the government to fulfil its housing policy objectives. The eclipse of local authorities as housing providers will increasingly draw private housebuilders into a corporatist relationship with central government and its agencies, the Housing Corporation and Scottish Homes, as suggested in Chapter 1. With regard to land use planning also, corporatism 'has, it seems, survived the ravages of recession and the anti-planning ideology of successive governments' (Saunders 1986, 134).

Lowe (1988) and Ball (1986,1988) have pointed to the significant emergence in the construction industry during the last twenty years of 'a handful of giant firms [which] have come to dominate what was previously a fragmented industry' (Ball 1986, 21). Most housing is provided by these firms, drawing on their extensive capital and land resources, to the extent that housing policy is now reliant upon their performance and cooperation. As Lowe (1988, 40) observes:

> Through the scale of their operations, their ownership of (or options over) extensive areas of developable land, and their ability to provide infrastructure and capital, the co-operation of the major building firms has become vital to the realisation of development plans and housing targets. Increasingly, they have come to enjoy a corporatist relationship with the planning system, mediated in particular through the Department of the Environment.

The influence of developers in the policy-making process can be seen in the introduction of joint land availability studies and the increased allocations of land being imposed on local authorities through structure plan procedures. The housebuilders want more (but not too much) land released. 'They seek a pliant planning system, not a deregulated one' (Lowe 1988, 46). Their major criticism of the planning process is that it is inflexible and they argue that land should be released in appropriate locations where people wish to live (HBF 1988). This is most clearly in the countryside, where the fastest growing counties in England are to be found in Berkshire, Buckinghamshire, Dorset and Wiltshire.

The other main political force acting upon the central state in its formulation of land use policy has been the ex-urban professional and managerial middle classes, who tend to be Conservative voters. As Lowe (1988) acknowledges, 'a central element of that concern has been to protect residential and recreational amenity', as suggested in Chapter 4.

> No longer is this restricted to green belts and outer metropolitan areas. Successive waves of middle class newcomers have spread out over the lowland countryside. Commuter hinterlands have vastly increased, pushed outwards by the regionalisation of urban housing markets under the pressure of high house prices...The greatest pressures have been experienced by accessible country towns and villages and picturesque countryside, with retirement migration going into the more remote rural areas.
>
> (Lowe 1988, 41)

Lowe explains how earlier ex-urban migrants have come to dominate the social and political institutions of the countryside, and he points to the local amenity and conservation groups which 'are particularly thick on the ground in the South of England and East Anglia, the areas which have experienced the greatest pressures for development; and in retirement areas', such as the West Country, the Isle of Wight, Cumbria and North Yorkshire. To the extent that these groups have developed a close liaison with the state, this has been developed at local level, however, rather than with the central state. For this reason, the contradictions between the corporatist influence of the volume builders on the central state and the anti-growth lobbying of ex-urban homeowners at local level have been reflected in an increasing tension between central and local government over housing land release.

Increasingly in the land use planning and residential development process, central government is exerting more pressure on local authorities through direct intervention (Reade 1987). Conflict is increasing over rates of land release and the tendency towards 'planning by appeal' (Herington 1987). Chapters 6 and 7 have already highlighted the impotence of local authorities in attempting to mediate between the central state and local populations. 'The local state has been responsive to the demands of the non-growth lobby, but given the structure of the land-use planning system has had to accept central directives, albeit under protest' (Short *et al.* 1987, 41). The main forms of control have been through planning appeals and structure plan alterations.

The ability of planning authorities to resist absolute growth in their districts is diminished by the power of the DoE to overturn refusal decisions at appeal. Herington (1987, 7) summarises the position of the planning authorities:

> In practice, local government has little choice to bargain with

> central government if it dislikes the extra growth because if it refuses to make additional land allocations in plans, developers are likely to put in planning applications which if refused will simply be allowed on appeal to the DoE. Thus counties have a dilemma: either identify more land in plans or face more development on appeal. Many would choose the former.

Alternatively, local authorities may prefer to refuse consent, at least to keep up anti-growth appearances, and blame the DoE for the ensuing development allowed on appeal. Indeed appeals have increased in the 1980s (the number in 1987 was 22,000 compared to 14,000 in 1983) and the proportion of appeals succeeding has increased over the past decade from 25 per cent to 41 per cent (Herington 1987).

Modifications to structure plan allocations by the Secretary of State have also reduced the ability of the local authority to carry out their anti-growth policies. Planning authorities have been criticised by housebuilders for objecting to increased housing provision in structure plan alterations on the grounds that it is merely a delaying tactic. However, the district councils may be successful in deflecting the imposed allocations elsewhere. Short *et al.* (1987, 40) cite the case of Bracknell and Wokingham District Councils to illustrate that since 'development often takes the line of least resistance...[the] intensity of local opposition can influence the location of new housing'.

As local authorities are increasingly unable to implement their anti-growth policies, planning by agreement or planning gain is becoming more common (Reade 1987). The planning authority may take the view of development that, 'if it cannot be refused it must be planned' (Short *et al.* 1987, 49), and aim to control the details of the development.

> They negotiate and bargain with developers 'privately', in an effort to ensure that in return for the granting of planning permission the developers provide, or meet the costs of providing, such things as landscaped public amenity areas, roads and services, or even residential or other development which was not the subject of the original application.
>
> (Reade 1987, 91)

The demand of housebuilders to build in areas of high demand, 'gives planning authorities a strong position from which to bargain with developers, even if they do not possess the authority to prevent growth' (Short *et al.* 1987, 41).

This authority is likely to be weakened further in the future if the government's proposals for *The Future of Development Plans* (DoE 1987) are implemented. This consultation paper proposes the elimination of structure plans prepared by county and regional councils.

District councils would continue to produce local plans, which would become more important as a result, but counties would produce only 'county planning statements' inevitably weakening the strategic element in land use planning. There must be a danger that district councils would seek to divert growth elsewhere, and that the only strategic control over this would be exerted by central government, thus leading both to greater uncertainty and to a further centralisation of power. Planning by appeal might become even more the established norm for resolving these issues, especially given the poor coverage of local plans in England. New housebuilding is likely to seek the line of least resistance, and

> in the absence of a coherent national and regional strategy for housing provision, incorporating an urban/rural dimension, development will tend to be 'extruded' between areas covered by the more robust area designations and/or protected by the most able planning authorities.
>
> (Hooper, Finch and Rogers 1988, 33)

Given the political pressures at central state and local state levels, and the tension, and indeed opposition, between these tiers of government, it is now necessary to ask what policies are likely to emerge, and what might be the implications both for different social groups seeking access to housing, for conflicting domestic property classes, and for the landscape and amenity of the countryside. These implications are now elaborated under three policy scenarios for increased land release, focusing in particular on the distributional housing market effects.

The first scenario then is one of a general relaxation of development control, which could involve village infill schemes, more widespread scattered development, village expansion or new settlements. Because several recent proposals for development have suggested the adoption of the 'new village' plan, these are considered as a separate, second scenario. This scenario looks at the adoption of both small and large scale new settlements. The third scenario is based on the development of housing for rural residents with low incomes, in the form of 'social housing', based on the recognition that there is a need for low cost housing for rent or sale that people in rural areas can afford.

Three scenarios

A general relaxation of development control

This scenario envisages not a 'developers' free-for-all', but certainly an increase in the release of land for private housebuilding. A general release of land might, in principle, allow several diverse forms of residential development in the countryside: the process of infilling existing towns and villages would no doubt continue. In addition more scattered dwellings might be built in the open countryside, estate additions or ribbon development might occur on the edges of existing settlements, or building might be concentrated in nucleated, free-standing, new settlements. The exact form will depend on the actions of the state (central and local), landowners and developers.

Previous settlement planning policies favoured nucleated developments, based on the assumptions that these were cheaper to service and less environmentally damaging than scattered and dispersed residential development.These assumptions, and their theoretical underpinnings, have been seriously undermined (cf. Cloke 1983, 168–75). Furthermore, the impact of these 'key settlement' policies has been shown to be detrimental to non-key settlements (Martin and Voorhees Associates 1980; ADC 1986). By concentrating development of both housing and services in certain locations, undesignated settlements and communities become less viable and depopulated, and hence one community is favoured at the expense of another. Concentrating further development in this nucleated form would be likely to continue the process of decline in undesignated villages. On the other hand, additional infill housing or additions onto the edges of such villages would mitigate these effects and possibly allow the retention of those services which remain.

Those villages which have previously experienced development and where services and transport links are already stretched, might benefit from a diversion of development away from them. Many villages are experiencing a disruption and strain on existing services, shops and roads brought about by the concentration of facilities within them and the additional role they have to play in catering for non-key settlements. However, while other governments (for example, in Sweden and Ireland) are questioning the assumed virtues of nucleated settlements (Jennings 1987; Folkesdotter 1987), there is little sign that either the central or local state in Britain is likely to depart from their long-standing aversion to sporadic development, even where conversion of farm buildings is concerned (Watkins and Winter 1988).

The attitudes of landowners to the release of more land for residential development under this scenario will depend on their motives of ownership. The smaller, petit bourgeois farmers with mostly non-pecuniary and expressive reasons for farming are most likely to be against selling land for residential development, unless for a clear local benefit, particularly in a scattered form because of the possible adverse effects on their farming enterprises (Shucksmith and Watkins 1988b). The NFU has highlighted these possible detrimental effects on farming, suggesting that 'the invasion of farmland by housing can have adverse effects equivalent to perhaps double the acreage actually lost' (Ambrose

1986, 187). Clearly such external costs will be minimised where development is most compact.

Farmers with high levels of indebtedness, normally the large scale capitalist farmers, are most likely to wish to sell their land to developers in order to obtain a capital gain, and this could become an increasingly important factor in landowners' attitudes to land release. Barlow (1988, 117) notes that 'many farms will become increasingly tied indirectly to financial capital – through mortgage indebtedness as land values fall – and to large chemical and pharmaceutical companies through biotechnology packages', and this may be expected to reinforce their market orientation. So, while larger capitalist farmers have generally opposed housing development in the abstract, as noted previously, their material interest will be quite different as potential vendors of land.

The suggestion has been made that farmers in lowland areas will have more opportunity to sell land than farmers in more remote areas, leading to 'the possibility of divisions between the "haves" – farmers with potential development land and profitable arable operations – and the "have nots" on the margin' (Barlow 1988, 117). This takes little account, however, of the considerable suppressed demand for housing, which might be made effective if cheaper houses were built, in both lowland and remote areas. Equally, a growth in households is projected by the DoE throughout Britain (see Figure 8.1). The growth in demand for retirement and second homes will further stimulate the demand for housing outside of the main employment areas, as the previous growth in demand for this type of housing in Wales, Scotland and upland areas of England illustrates. Hence, the division between landowning 'haves' and 'have-nots' may not be regionally based, as Barlow (1988) suggests. More significant may be the question of which farmers will wish to sell land and for whom. The third scenario explores the possi- bility of landowners releasing land for meeting local housing need, in line with the NACRT approach.

The theoretical arguments which suggest that planning controls have increased the price of housing (reviewed in Chapter 3) similarly suggest that a general relaxation of planning restrictions would reduce real house prices, or at least mitigate their increase. However, Barlow (1988) suggests the very reverse: he argues that landowners are likely to act oligopolistically against developers so that 'the reality of deregulation may well be generally rising land prices, at least in desirable locations, and a probable leapfrogging by developers over areas of powerful landowner alliances' (p. 114). There is, however, no evidence that landowners are likely to act in such a way. Munton and Goodchild (1985) were highly sceptical of the suggestion that any individual landowner can exert monopoly power over land to force up prices, and they found no instance of landowners consciously colluding to achieve such an effect: 'owners infrequently act together and the land held by one owner can usually be substituted for land held by another' (p. 180–1). Thus, 'development land prices will not rise in real terms come what may unless the supply of development land is continuously and increasingly squeezed by public policies below the amount demanded' (p. 180). A scenario of more liberal planning controls is therefore likely to be characterised by falling, not rising, real house prices, contrary to Barlow's assertion.

The related distributional issue, then, is whether the decrease in the real price of housing in rural areas would be sufficient to allow low-income households to obtain access to owner-occupation. At present, nearly half of all households in the south of England earn too little to buy a house (Bramley and Paice 1987), and the even greater polarisation of income levels in rural areas (McLaughlin 1986) is such that a simple relaxation of planning controls is unlikely to be sufficient to meet the needs of lower-income groups. Because the level of rural house prices tends to relate to the income level of the non-manual (higher-income) households, who exhibit effective demand, house prices are even higher in relation to low-income households than in urban areas, so that 'the poor in the countryside are even less well equipped to compete with their wealthier neighbours [for home ownership] than the poor in the cities' (Platt 1987). Rural house prices are likely to remain high given the demand from wealthy incomers as well as from high income 'locals', as already noted. The presence, or potential presence, of these components of demand acts not only to keep rural house prices at a high level: in addition, it casts doubt on any strategy of attempting to meet the rural housing needs of low-income groups through merely relaxing planning controls. It is clear that any additional stock at slightly lower prices would predominantly be sold to ex-urban migrants and commuters and not to lower-income households, as higher-income ex-urbanites would outbid those in rural manual occupations for as many new houses as were built, unless a massive tide of development was unleashed.

The presence of high-income groups in owner-occupied housing in rural areas, in conflict with non-owners over the release of land for building, might act to dampen the fall in house prices in any case. Through their opposition to development proposals in their locality, local shortages of building land might be engineered, leading to an enclave of higher house prices and reinforcing the locality's social status. These effects are likely to be associated with the protective designation of areas of countryside as green belt, AONB, national park, etc., which are therefore likely to experience the most intense gentrification.

Although a small fall in house prices is likely to occur under this scenario, a general relaxation of planning controls is likely to benefit only middle-income groups who would thereby gain access to home ownership in rural areas. Higher-income groups would still be able to outbid poorer households for the available stock as it is unlikely that the stock would increase sufficiently to cater both for non-manual and professional groups and for poorer manual workers. Private developments would cater for professional and non-manual workers, to the exclusion of young people, old people, the unemployed and others with insufficient market power.

> Speculative housing could reduce social polarisation in a relative sense by allowing a greater number of the 'middle bands' of the [urban?] workforce to gain access to cheaper housing...but the problem of housing need would not necessarily be touched.
>
> (Herington 1984, 46)

The beneficial effect on the average level of house prices in increasing access for middle-income groups must be set against the losses, for example, to the countryside and green belt; and this general increase in the release of land for private housebuilding does little to relieve the continued disadvantage of lower-income groups, either in their access to housing or in their ability to share in rising domestic property values.

Growth concentrated in new settlements

This scenario envisages the development of new free-standing 'villages', alongside a policy of growth prevention of existing towns. Hence, village infill and expansion would be diverted to these new 'villages'. Several 'new villages' have already been proposed: about twenty such villages have been mooted in and around the London green belt by Consortium Developments, an alliance of the volume housebuilders, and a group of farmers outside York is proposing a mega-village for 12,000 people.

The central feature of this scenario is that development pressure is diverted away from the wider countryside and concentrated in relatively few locations. Existing villagers, relieved of the pressures of infill in particular, would benefit from environmental improvements and would live at a lower density. The DoE clearly supports new villages as a preferred alternative to sporadic development, and although it accepts infilling, a recent consultative paper emphasises that too much would be detrimental to the character of villages: 'it would be a mistake to pack new development in too tightly when there is no need to do so and when there are better alternatives available including new settlements' (DoE 1988a, 11).

New settlements would no doubt suit the volume builders far better than piecemeal schemes or sporadic development, since they will be able to outbid all other purchasers for the large blocks of land involved, and then determine themselves the pace at which the land is developed and houses released onto the market, so giving them the maximum possible scope for profit maximisation. The central state's enthusiasm for new villages is likely to be bolstered, therefore, by the HBF's corporatist support.

Despite the generally assumed coincidence of minimal service costs with nucleated settlement forms, the idea of new settlements has been criticised on the grounds that this diversion of development would not make efficient use of the existing infrastructure, nor utilise the spare capacity of existing settlements: instead, both public and private investment much needed for inner-city redevelopment and renewal would be channelled away from the urban areas. The DoE has stressed that 'it is also essential that opportunities for increasing the supply of housing in rural areas should not detract from the policy of recycling urban land' (1988a, 14). Urban local authorities are therefore likely to oppose new villages in the hope of forcing investment into the inner city: Strathclyde Regional Council has made a particular virtue of such an approach, seeing its strong opposition to greenfield development as the major factor in channelling the private sector into the rejuvenation of the merchant city. In rural areas, however, the local state's attitude to new settlements may be ambivalent: against its normal anti-growth sentiments, a proposal for a new village may appear cost-efficient in terms of service provision (especially where infrastructure is to be provided at the expense of the developers), and offer, in addition, the opportunity to avoid a lengthy sequence of smaller disputes with anti-growth interests.

Clearly, new settlements are likely to engender hostility from those whose rural retreat will be transformed into a suburban setting, and whose house price is likely to suffer substantially as a result. 'To protect their green vistas, the Home Counties' middle-classes are banding into action committees with resonant, sinister-sounding acronyms such as Stamp, Sound and Save' (Lean 1989). Villagers away from these sites, however, will materially benefit from the diversion of development pressure away from their homes and environs, not only protecting their amenity and their status but also in all likelihood enhancing their property values. It will be interesting to observe whether those who proclaim their hostility to development in their own localities are as quick to oppose the construction of distant new villages, and so to eschew their material self-interest.

The question of farmers' and landowners' attitudes to land release may be less relevant to this scenario, since the state could conceivably assist the developer with compulsory purchase powers once a site is identified in plans. If the sale is voluntary, then again it may require agreement with perhaps only a single landowner to obtain sufficient land for a small settlement, and the potential gains are so large that most farmers and landowners will find them difficult to resist. Some farmers are even initiating their own new village proposals, and these tend to be the larger capitalist farmers with land to spare rather than small family farmers, for whom survival is paramount. However, it is unusual for the farmer or landowner to act also as the developer: generally 'landowners just sit back and enjoy their enormous unearned increments' (Ball 1986, 20).

The environmental impact of new villages is as controversial as that of the relaxation of development control. Developers argue that 'they can improve on scruffy and underused agricultural land by smartening it up, putting in new woods, restoring wild meadows, providing a habitat for wildlife' (*Guardian* 2 January 1988). For example, the plan for the new township of Brenthall Park, to be built east of Harlow, includes a country park and wooded area and only 3,500 dwellings with a proposed population of up to 9,000 people. Similarly, the proposal for Foxley Wood in northeast Hampshire included a community of about 12,500 people in 700 acres, of which forty acres will be open space and thirty two acres a continuous water park and the whole site will be enclosed by a thick, coniferous tree screen (Bennet 1987). The site proposed for Foxley Wood is on disused gravel workings, and to this extent the scheme will be an environmental improvement, although it will also intrude on an SSSI. Despite a favourable reaction to Consortium Developments' proposal for Foxley Wood from Nicholas Ridley, the scheme was rejected by his successor at the DoE, Chris Patten, partly to establish his 'green' credentials with Conservative voters. Other proposals, such as that at Stone Bassett in the Chilterns, are less attractive in environmental terms.

In considering whom the new settlements will provide for, it is important to consider the scale of provision. Very few of the existing proposals for new settlements are actually at a 'village' scale of 200–1,000 houses (DoE 1988a); most proposals in reality provide for new towns of about 12,000–15,000 people, covering a range of different house types from starter homes to large dwellings.

Since none of these proposed developments has yet been built, it is perhaps worthwhile to look at the experience of the new towns to illuminate the possible housing market effects, and to suggest the type of people likely to live in the new 'villages'. Research on the nature of migration to new towns has shown they have been very selective in the type of migrant they attract (Shaffer 1972; Heraud 1968). Many more professional and semi-skilled migrants were attracted than unskilled and unemployed (Sim 1983). If new villages were to provide only large private housing, they too could be dominated by high-income and professional groups, and lower-income rural workers would be unable to gain access to owner-occupation. However, if proposals include housing for both high- and low-income groups, with housing for rent and for sale, a more balanced housing provision would occur. Housing for low-income groups has been included in the plans of Consortium Developments for Foxley Wood and Stone Bassett, and in Countryside Properties' approved development at Brenthall Park. Brenthall Park, it is said, 'is planned to provide for a total community with a wide range of house types, as well as a broad tenure mix which will include housing for renting and various forms of shared ownership' (*Guardian* 2 January 1988). Similarly, the chairman of Consortium Developments stressed their commitment 'to the creation of a balanced community. New country towns like Foxley Wood are not just executive ghettoes by another name' (Bennet 1987). Indeed, in the unsuccessful proposal for Foxley Wood, the support of the North Housing Association was enlisted, with £60 million of private funding raised to finance a range of social housing as part of the 4,800 home development on 600 acres. If new country towns were to follow this form, then some contribution would be made to the provision of housing for low- income households and solving the rural housing problem. However, it is possible that the scale of the social housing provision, compared to that for more prosperous groups, may still be insufficient for the level of need. Also, given that such housing will be highly sought after, but not rationed by price, priorities for access to the social housing built will also have to be developed and monitored. A further problem is that many low-income groups may wish to remain in their own village rather than move to a new settlement some distance away.

A final point to note is that under such a scenario, boundary problems may well arise between authorities. One strategy for planning authorities, inclined to oppose growth, is to argue that no suitable sites for new country towns exist within their own area and to pass the burden of coping with growth onto a neighbouring district. Under such circumstances, or where new villages are genuinely inappropriate because of specially designated green belt or AONBs, for example, the authority will seek to ensure that adjoining schemes do take account of neighbouring demands and needs. For new villages to be successful, as a general policy response, an overall regional strategy must be developed to ensure that all rural dwellers have the same opportunity for moving to a new settlement.

Social housing developments

The two scenarios reviewed thus far suggest that a general relaxation of planning controls would be insufficient to help house lower-income groups in rural areas. Further, a concentration of growth in new villages, while perhaps incorporating rented housing and shared ownership schemes to house such groups, would nevertheless fail to address broader issues of rural development and social segregation. Low cost housing would not be available within existing rural localities, which would tend to lose all but their high income, professional and non-manual population as a result. Although under each of these scenarios there may be a beneficial effect in that the average level of real house prices in rural areas would decline, this would not be sufficient in itself to ensure the housing of lower-income groups in rural areas. A more sophisticated approach to the release of land for social housing is needed if this end is to be achieved.

Earlier chapters have established that those groups most disadvantaged in the rural housing market tend to be elderly people, and young people under thirty without children, often on low incomes and without secure accommodation. Such groups are always least likely to gain access to housing where the available houses are allocated according to ability to pay. While the release of more housing land and the development of new country towns may allow more households amongst the middle-income groups to gain access, the poorest will continue to be denied access to housing in rural areas unless some houses, at least, are allocated by another mechanism. Housing opportunities in rural areas for those with the least means can only arise through an alternative allocative criterion to ability to pay. What is required to achieve this end is the building of low cost housing in rural areas, affordable by those with the least means, and the allocation of that housing in such a way that it goes to meet the needs of such households. For many years the means of achieving this has seemed to lie with the provision of state subsidies, either in the form of revenue grant (e.g. housing support grant to local authorities) or of capital grants (e.g. HAG to housing associations). While the willingness of the government to offer such subsidies on the scale required is now in doubt, it must nevertheless be recognised that in virtually every other European country low cost or social housing is only possible with some element of tax relief or subsidy.

Following the statement by the DoE (1988a) on *Housing in Rural Areas* it seems clear that housing associations are seen by the central state, at present, as the most suitable means of providing such accommodation. It is relevant therefore to consider the recent experience of rural housing associations. It has already been seen in Chapter 7 how rural housing associations have been able to acquire cheap land with the support of some local planning authorities and a few philanthropic landowners, thus enabling them to pass the saving onto the future occupiers of low cost housing. But can such an approach constitute the central state's strategy for new housing in the countryside? For while this ad hoc arrangement appears to work well at present, it works only at a very small scale of provision, and only a few houses have been built in this way throughout rural England.

A slightly different approach to providing low cost housing in rural areas has been through the development of joint venture schemes in which private housebuilders provide low cost housing under licence on cheap local authority land and the local authority retains the right to nominate purchasers (Clark 1988). However, most have failed to keep control over subsequent resale and the first householders have made considerable profits as the house is resold at market value and is lost as an affordable house. This is not therefore an adequate long-term strategy, and is unlikely to attract support from landowners and the local community 'if it will result in a windfall for first purchasers' (Clark 1988, 60).

If a strategy for broadening access to rural housing is to be developed on the scale required to meet the needs of lower-income groups, then it will not be sufficient to rely on opportunistic ad hoc schemes, time-consuming searches and the goodwill of a few philanthropic landowners, as NACRT has to do at present. As the scale of social housing provision grows under this scenario, planners will become less willing to make exceptions because of the fear of precedent, and more landowners will anticipate development gains, so demanding a more general solution from the central state. To succeed at the necessary scale, such a solution must be built into planning legislation and thus into land values, rather than tacked on as an occasional ad hoc exception to planning policies.

The essence of the NACRT's approach is concerned with reducing the development value of the land, and hence the land cost to the developer, and with passing on that reduction to the consumer in the form of lower housing costs. This could perhaps be achieved on a more widespread scale through the creation of a new planning use class, through which some land would be permitted to be developed only for social housing (Shucksmith and Watkins 1988e). The effect of this would be to restrict the bidders for such land to housing associations and local authorities, and to place them in a position of virtual monopoly. Owners of land zoned for social housing would either have to accept the (low) price offered by the housing association, or hold the land in the hope that a later development plan review or a planning appeal would confer on them the full development value. At some price above the agricultural use value, but below the full development value, one would expect that landowners would be willing to sell to housing associations.

Indeed, small, petit-bourgeois farmers might actually prefer to sell to housing associations on such a basis. Despite their general reluctance to sell land, family farmers may be favourably disposed towards the release of small amounts of land for housing which they believe will benefit local people, in accordance with their ideology of localism, kinship links and the rural 'moral economy' (Shucksmith and Watkins 1988b). Thus, Newby *et al.* (1981) found that in Suffolk, small farmers consistently supported council building in the remoter areas, contrary to Barlow's (1988) view that family farmers oppose new developments. On the other hand, the Country Landowners' Association (CLA), while 'committed to finding ways to encourage the release of land to suitable agencies for the provision of social housing', has observed that 'even the most philanthropic landowner can hardly be expected to present it [land which would obtain full development value] to a rural housing association' (Dunipace 1988, 55).

Crucially, the greater the certainty created by local and central government that full residential consent would not be forthcoming, the lower the price of land zoned for social housing would be, since 'hope value' would be reduced. If landowners' expectations of development gains were effectively dampened on land zoned for social housing in this way, then rural housing associations would expect to be able to purchase land cheaply and on a scale sufficient to address rural housing needs. Their development costs would be reduced substantially, so enabling them as non-profit providers to offer low cost housing throughout rural Britain either to rent or on a shared equity (non-staircasing) basis. Clearly substantially more capital would be required for an expanded programme, and the DoE (1988b) notes that the Housing Corporation will be providing considerably increased funding over the next three years for rural housing association schemes, albeit only sufficient to build 600 homes a year, whilst 'the Rural Development Commission plans to increase its funding of NACRT, to up to £315,000 in 1988/89', with the possibility of further increases in subsequent years. The provisions contained in the Housing Act 1988 also include the involvement of private sector finance in the provision of low cost housing in rural areas, although as discussed in Chapter 7 this has largely been restricted to the provision of shared ownership housing in rural England. The attraction of private finance into the provision of low cost housing in rural areas is not likely to be successful without a specific tax concession for private investment in rural housing asssociation schemes. Such a tax relief would not only attract capital into the provision of social housing in rural areas, but the availability of cheap credit would also further reduce the costs of provision and ultimately reduce the cost to the consumer. Cheap credit, combined with cheap land, would allow genuinely affordable housing to be built in rural areas for those rural households who at present have no prospect of finding housing in the countryside.

The numbers wishing to purchase or rent such houses at low cost will clearly be very great, so that the homes built would have to be allocated according to criteria other than ability to pay; such as housing need or contribution to countryside policy objectives. In addition, a sufficient amount of land would have to be zoned for social housing in order for the programme of building to keep pace with needs. In most areas one might expect that this form of land release would be additional to land release for private development, while in areas of particular landscape value (such as the Lake District) land might only be released for social housing; but this balance would be a matter for local and central planning authorities to determine, in collaboration with the Housing Corporation or other funding agency.

A slightly different approach for areas where there were no overriding reasons for restraint on housebuilding, might be for local planning authorities to permit joint venture schemes to be built on land zoned for social housing, as proposed by Middleton (1988). The cheaper cost of land would allow the developer to obtain additional development profit on the houses built for open market sale and this would be used to finance the construction of social housing for rent, as part of a planning agreement. Middleton (1988) suggests that 'community need housing sites' should be included in planning policies (where 'community need' is defined as 'the need of a community or village to have a sufficiently varied housing stock to cater for all age ranges and size of household units'), and could contain both private housing and special provision. 'For example, a community need site could have 10 small units – 7 for private sale, 3 owned by a housing association for "local need" cases.' Such schemes would not be dependent upon government funds, and nor would any public sector borrowing be implied. However, such schemes might not be suitable where planning authorities believed it appropriate to release only a small amount of land for housing, perhaps because of agricultural, landscape or wildlife considerations. In these more constrained circumstances, capital would have to be borrowed to finance schemes of pure social housing. Because of local variations in housing need and in planning circumstances, each of these approaches might be appropriate within a single district.

Within the most pressured commuting areas around London and the southeast, there may also be some areas where more housebuilding is acceptable while other areas, such as green belt and AONBs, are unsuitable for large-scale construction. An appropriate strategy to meet both regional and village housing needs might therefore combine the release of land adjoining existing villages for social housing only, especially in green belt areas, with, at the same time, a concentration of large-scale private development to meet regional needs in a few 'new villages' in less sensitive countryside. Inevitably this would lead to some degree of tenure polarisation in existing villages and to spatial segregation of upper- and middle-income groups between existing and 'new' villages: but arguably this would be less serious than the continuation of current trends towards the total exclusion of lower-income groups from rural areas. Some social polarisation will result but it will be less than that which would result from a failure to construct social housing in rural areas.

It is worth noting two indirect effects of such a strategy. First, the effect on existing rural residents' house prices might be broadly neutral, or even positive if the decline in amenity value and social exclusivity caused by building social housing in villages is offset by the increase in price consequent upon a tighter constraint on private housing development. Thus there would be no necessity to depress private house prices in order to broaden access to rural housing. Second, another effect would be a major increase in rural construction activity, giving a substantial and highly cost-effective boost to rural employment and the rural economy in general, as well as benefiting the volume housebuilders.

Such a scenario may therefore allow the housing needs of low-income households to be met in rural areas: but there is no escaping the fact that the feasibility of the strategy hinges on the attitudes of farmers and landowners and their willingness to sell sites which will assist local people to be housed, perhaps in accordance with a rural 'moral economy', and to accept moderate development gains rather than hold out for the possibility of much larger development gains some years later. If only family farmers are willing to sell, this may affect the pattern of social housing provision, and its spatial equity, leaving larger farmers either richer or with no gain at all according to subsequent land release policies for private housing. Farmers' propensity to acquiesce in such a strategy can only be established by empirical enquiry.

Conclusion

Despite the success of urban renewal policies and government initiatives for the development of vacant or underused urban land, it is clear that national and regional housing requirements far outreach the possible supply of brownfield sites. Therefore there will be a continuing requirement for greenfield sites for private housebuilding. How this land is released has significant implications for who will benefit from increased access to housing, and from changes in property values.

A discussion of the forces acting upon policy-makers, and of the changing corporatist liaisons of the central state, suggested that the volume housebuilders have gained in influence now that their cooperation is essential to the success of housing policy. While the central state is increasingly open to private developers' influence, however, the local state is thought to reflect the anti-growth interests of home-owning rural residents, bringing central and local government into conflict through the planning system. It seems that this may lead to further centralisation of power, as local planning authorities seek to refuse or divert development, and the central state intervenes through planning appeals and structure plan modifications. Whether this continues on an ad hoc basis, or proceeds according to a centrally imposed strategy, remains to be seen. Again, this will be important in determining which social groups gain and lose.

The distributional effects of possible changes in planning policies, stimulated by the current land use debate, have been examined under three main scenarios of potential housing development. A general relaxation of development control is unlikely to increase access to rural housing for lower-income groups since such households will continue to be outbid for housing by more affluent groups, even if real house prices fall. The development of new country towns is also thought to be insufficient to solve the rural housing problem, although these could make a useful contribution to housing the projected regional and national increase in households over the next decade. The provison of affordable housing at a suitable scale to meet the needs of low-income rural dwellers might be assisted, however, by the introduction of a policy of zoning land exclusively for social housing. In most areas one might expect that this form of land release would be additional to land release for private development, while in areas of particular landscape value land might be released only for social housing. Even within the most pressured commuting areas around London and the southeast, there may be some areas where more housebuilding is acceptable while other areas, such as green belt and AONBs, are unsuitable for large-scale construction.

An appropriate strategy to meet both regional and village housing needs might therefore combine the release of land adjoining existing villages for social housing only, especially in green belt areas, with, at the same time, a concentration of large-scale private development in a few 'new villages' in less sensitive countryside. Inevitably this would lead to some degree of tenure polarisation in existing villages and some spatial segregation between existing and 'new villages': but arguably this would be less serious than the continuation of current trends towards the total exclusion of lower-income groups from rural areas. Some social polarisation will result but it will be less than that which would result from a failure to construct social housing in rural areas.

Finally, a strategy of zoning land for social housing, and thereby tending to reduce its development value, has been shown to depend crucially on the attitudes and willingness to sell of farmers and landowners. This chapter has suggested that some farmers might be expected to sell land for social housing to meet local needs, in accordance with an ideology of localism and a rural 'moral economy', and if this proves correct then a new social housing use class might prove successful. This suggestion now requires empirical testing before it can form the basis for any policy advice.

Chapter nine

Conclusion

This book set out to address the twin themes of the conflict of public policy objectives relating to the competition for land between landscape preservation and rural housing provision, and of differential access to rural housing among competing groups, and how these distributional outcomes are modified by policy. In addressing these two themes, perhaps inevitably, other concerns have also emerged and avenues for further research have become apparent. It is the purpose of this concluding chapter to attempt to pull together these various strands, offering some conclusions relevant to the objectives of this book, in both instrumental and methodological terms. This should help to inform future research into rural housing markets, both in terms of empirical concerns and in terms of methodology, as well as perhaps assisting in the formulation of policy.

The pattern of rural housing disadvantage

The discussion in Chapter 5, as well as the case study, elaborated the differential access to housing among competing groups, and the role of public policy in helping to structure this pattern of advantage and disadvantage. Chapter 5 indicated that access to owner-occupation is most difficult for households with low incomes and little wealth, especially if they are elderly, in manual occupations or unemployed. Access to the declining private rented sector is largely restricted to farm and estate workers who may be offered the tenancy of tied cottages. Access to council housing is hardest for young single people and for elderly applicants, although many 'general needs' households may have to wait a considerable time while the few vacancies arising are allocated to priority groups such as homeless persons, evicted farmworkers (in England) and special needs groups. Since most housing in rural areas is owner-occupied and not rented, the main criterion of access to rural housing is ability to pay rather than housing need, local ties of kinship, or potential contribution to landscape management. As Newby (1980,

179) has noted, this is significant 'in moulding the social composition of the rural village', and it may eventually also have a more visual effect on the quality of the landscape, as argued in Chapter 3.

The economic basis of housing disadvantage was also apparent from the case study. In the Lake District low-income and even middle-income households were denied entry to the market for owner-occupied housing by high house prices which reflected not only the area's attractive character but also the existence of policies designed to maintain that character. Again, there was very little rented accommodation available, and the few council lets tended to be allocated to families and pensioners, leaving childless couples and young single people as the most disadvantaged households. In other localities, such as Uist (Shucksmith 1990), while the mechanisms for housing allocation may be quite different, the economic basis of housing advantage and disadvantage appears to remain the same. These findings concerning the economic basis of housing disadvantage suggest that one area for future research might involve an elaboration of the relationship between labour markets and housing markets, and of the extent to which housing market situations are dependent on production relations. The need for further empirical investigation of the relationship between domestic property classes and acquisition classes has already been noted in Chapter 4, and this is discussed further below.

It has been noted in the case study, as well as in Chapter 5, that housing disadvantage is clearly related to stage in the life-cycle, with the young and the old more constrained in their housing choices than families of middle years. This relationship may also be worthy of further exploration, perhaps through studies of households' housing histories. Such work has been undertaken in Sweden by Thunwall (1987), who has found evidence of distinct housing requirements at different stages in the family life-cycle, associated with local migration: in the British context it would be relevant to ask whether such a relationship also exists and to assess the social consequences of a failure to meet, say, the housing requirements of young people in rural areas. This concern with life-cycle aspects may also contribute towards a refinement of policies, as for example in crofting areas where it was suggested that improvement policies might be more successful if they were more sympathetically adapted to the circumstances of the elderly crofters who typically live in BTS crofter housing.

While constraints on housing choice clearly relate to the economic circumstances and life-cycle stage of the household, these have not been the dimensions of most interest to previous researchers and to policy-makers: instead they have been more concerned with the relative housing market opportunities of locals and non-locals and with the merits of local claims to rural housing (cf. Rogers 1985a), as discussed in Chapters 3 and 4 and in the Lake District case study. It has been suggested in this book that the concept of local needs relates less to housing choice and more to the politically acceptable rate of housing development in a rural area – a point discussed further below. Nevertheless, in the context of identifying patterns of disadvantage, it is apparent that while the majority of houses in a rural area are allocated according to ability to pay, households reliant on an income from typically low paid occupations in the locality, such as agriculture, forestry and tourism, will necessarily be disadvantaged relative to better paid non-manual workers, whether they live or work in the locality or elsewhere. To this extent, many (low paid) locals may be disadvantaged (although other high paid locals will not be). The implication is that a non-market basis of house allocation may be a prerequisite of any successful attempt to discriminate in favour of young, low paid locals, if such discrimination is considered desirable.

Finally on this theme, it has been suggested that the construction of a typology of housing consumption classes may be helpful in elaborating the pattern of housing disadvantage, despite some theoretical misgivings about whether these can really be termed classes or if they are more properly housing status groups. In Chapter 4, Saunders (1984) was cited as arguing that consumption cleavages are, in principle, as important as class divisions in understanding the stratification of society, and that housing consumption cleavages in particular play a key role in affecting life chances and modifying economic inequality. Such consumption cleavages were investigated in Chapter 5 and in the case study, as patterns of housing disadvantage were inferred from analyses of house allocation mechanisms in both public and private sectors. However, it may also be worthwhile to attempt to derive empirically a more formal typology of households, using a cluster analysis of household survey responses (cf. Shucksmith 1987): such a survey-based method might be tried in future studies and in different contexts to assess further its general utility.

The modification of rural housing disadvantage

Of central importance to this book has been the assessment of how the pattern of rural housing disadvantage has been modified by public intervention, either deliberately or incidentally. Both housing policies, intended to ensure a decent home for all, and planning policies, charged with balancing the competing claims of housing construction and landscape protection, have been analysed, and the conclusion has been reached that, apart from those groups fortunate enough to have gained access to the few council houses built in rural areas, it is generally the wealthy, more privileged groups in rural society which have benefited from public policies.

As Hall *et al.*(1973), Rogers (1976) and Newby (1980) have pointed out, post-war planning policy in Britain has actively discouraged rural housing construction, with socially regressive and socially exclusive consequences. The justification for such a policy has rarely been questioned, except in northern Scotland where housing has been seen to act as a constraint on rural development (Thomson 1982) and in the southeast of England where the pressures of urban growth have become focused on the issue of housing land availability. However, such policies are being reassessed in Ireland and Sweden, among other countries, and there may be merit in querying whether they continue to be appropriate to Britain's rural areas. Clearly, the immediate post-war imperative of maximising food production, and therefore, cultivating every possible acre, is less powerful in the context of EC surplus production, dairy quotas, price cuts and land set-aside. Further, the economic justification for the concentration of service provision has been called into question both by Gilder's (1979) empirical findings – notwithstanding the methodological doubts expressed by Cloke and Woodward (1981) – and also by Jennings (1987) who argues that higher costs in service provision are largely a matter for the pricing policies of public utilities. The Scott Report's proclamation of the social benefits of domestic proximity appears to be at odds with many people's desire to live at lower densities in the countryside (Jones 1987). It seems therefore, that the sole firm justification remaining for blanket discouragement of rural house construction is the damage to the landscape which additional houses would entail. This is clear from the then minister's statement that the government's revised planning advice of February 1987 'does not mean that there is no constraint on the development of other [non-prime] agricultural land; what we now stress is the protection of the countryside for its own sake rather than for the productive value of the land' (Waldegrave 1987, 10).

The conflict of objectives relating to the competition for land between house construction and landscape protection is therefore inextricably linked to the distributional theme of differential access to housing among competing groups. However, the argument rehearsed in Chapter 3, that landscape protection ultimately rests on the maintenance of viable rural communities (and therefore on housing availability among other things), is gaining currency (MacEwen and Sinclair 1983; World Conservation Strategy 1980). This presents local authorities and central government with a series of dilemmas, of which the desire to resolve the competing claims of housing and amenity protection is only one. The case study of the Lake District reviewed an attempt to resolve these claims through limiting new house construction to those built for local occupancy only, but this was found to be unsuccessful: not only did this device fail to help the local households identified as disadvantaged, but its distributional effects were regressive. Chapters 7 and 8 discussed a number of alternative mechanisms, and several appropriate policy responses were suggested, including notably that of designating certain land for social housing only: however, these avenues were found to be prejudiced by the imposition of central government constraints on housing investment.

In more detail, it was suggested that where landscape quality dictates that only limited new building is desirable, the local planning authority might refuse permission for all private housing and instead reserve the limited supply of available land for rented social housing, built by local housing authorities and housing associations. This might be combined with increased municipalisation of the existing stock, and with short-leasing schemes. Such a strategy is not feasible at present, however, because local planning authorities are prevented by the Secretary of State from discriminating between private housing and social housing (except in exceptional cases where an adequate provision for private builders has already been made), and also because centrally imposed restrictions on borrowing prevent local housing authorities and housing associations from building sufficient social housing.

Even where landscape quality is less critical, it was argued that merely permitting private speculative developments would be unlikely to assist disadvantaged groups significantly, because they would continue to be outbid in an open market. Social housing provision and widespread self-build initiatives would be more helpful to such groups, but these were currently precluded again by central government constraints on investment. Self-build schemes in particular appeared to be worthy of further research, perhaps focusing at first on comparative overseas experience. Such schemes were considered highly appropriate to rural areas, both in the provision of dispersed housing on small sites and in utilising the skills possessed by manual workers in rural areas, but their potential for more widespread application was hindered both by the difficulty of acquiring sites and by the lack of cheap credit for lower-income groups embarking on such projects. Again, any rural housing strategy which sought to promote widespread self-build by lower-income groups would require collaboration between local and central government, and this seems unlikely to be achieved.

The conclusions drawn from Chapter 7 and from the case study strongly suggest the impotence of local authorities and other local 'managers' in attempts to mediate between central government and the local population. In the Lake District, especially, the dilemma of the local planning authority, the LDSPB, was acute. On the one hand they were directed by section 37 of the Countryside Act 1968 to have regard to what they saw as a social priority: the housing needs of 'the young people on whom the future of the community depends' (Feist *et al.* 1976, 95). On the other hand, the government had stated that their primary duty was to preserve and enhance the landscape, and they faced the opposition of the Secretary of State to any attempt to resolve this conflict of objectives. In any event, they lacked the powers to intervene in an effective way. Shaw (1980) regarded their frustration in having identified the problem but being powerless to resolve it, as symbolic of the impotence of local authorities. A similar frustration is apparent in the housing plans of many local housing authorities in rural areas, who propose programmes of investment to meet identified housing needs in the knowledge that they will be prevented from implementing those programmes by central government's imposition of limits on their borrowing.

This impotence of local government tends to confirm Pahl's (1977) view of local managers as middlemen performing a mediating role between central state authority and the local population, subject to major constraints imposed by central government. In Pahl's own research, he complemented his analysis of local authorities' mediating role with analyses both of the autonomous actions of the central state and of the private sector's activities, and developed a theory of corporatism to address these issues. The conclusions of this book concerning the impotence of local authorities and the constraints imposed by central government upon their intervention in rural housing suggests the value of future research which focuses on and elaborates the process of central government policy formulation in relation to rural housing. How far, for example, have central government's policies affecting rural housing derived from

> a division of labour within the government and the Civil Service which has allowed the Ministry of Agriculture to respond to a clear and unambiguous demand to reduce the cost of food production, while other departments, both locally and nationally, have been left to mop up the social consequences'.
>
> (Newby 1980, 264)

Cloke (1987) considers the influence of the central state on rural planning to have been enormous, but recognises that the nature of this influence may be viewed quite differently according to one's theoretical perspective. While some researchers have seen policy formulation by central government to represent the dominance of a powerful elite, or indeed as merely subordinate to the interests of private capital, others have adopted a pluralist position in which the interplay of pressure groups and other political forces have a strong influence on policy. Rocke (1987, 180) believes that central government acts increasingly 'as a corporate sector' in league with private capital, weakening planning controls, to ensure the realisation of development potential. On the other hand, researchers concentrating on the role of pressure groups, such as the CPRE, in thwarting the relaxation of planning controls naturally tend to subscribe to a pluralist position in which the relative strength of the various lobbies concerned with rural housing policies is gauged. Short *et al.* (1987) follow Saunders (1981) in adopting a more complex view in which the state

> is neither neutral nor a single body, and the conflict between housebuilders and community groups through the planning system produces conflict between central and local government.
>
> (Short *et al.* 1987, 29)

> The contradictory functions of social investment and consumption have progressively come to be located at different levels of the state [central and local government, respectively]. In addition, the mode of policy development and implementation is markedly different. In many cases, social investment is determined within a corporate sphere of interest mediation whereas consumption issues lie more within the open sphere of local politics.
>
> (Short *et al.* 1987, 39)

In Berkshire, therefore, they concluded that

> The local state has been responsive to the demands of the non-growth lobby, but given the structure of the land use planning system has had to accept central directives, albeit under protest....The central state is concerned with macro-scale issues and growth at the national level. The continued vitality of buoyant areas such as Central Berkshire is perceived to be threatened by the anti-development stances of the local planning authorities. However, the Conservative government is keenly aware of the extent and strength of opposition to widespread residential growth within its own political heartland.
>
> (Short *et al.* 1987, 41)

Whether such conclusions are more generally applicable to areas such as national parks and peripheral rural areas is doubtful, but this is not the point. What is significant is the attempt to unravel the forces weighing on central government as it determines its policy on rural housebuilding in one case study area. While this book has deliberately adopted a partial scope which excluded the process of policy formulation in central government, the findings of local government impotence and of the dominance of central government suggest that further studies of this type, from a variety of theoretical perspectives, should be conducted in order to identify the forces which underlie central government policies and govern the constraints on local authorities.

Further work is also necessary at the local level, particularly with respect to the rural application of Saunders' concept of domestic property classes. The discussion in Chapter 4 suggested that this concept might illuminate the more fundamental class basis of conflicting policy objectives, policy formation and distributional outcomes. Empirical work is required to investigate whether local opposition to new housing arises from the vested interests of home owners, protecting the accumulative potential of their properties, or merely from a common interest in preserving lifestyles. Are anti-growth interests at root production or consumption interests? It is also necessary to elaborate further and refine the concept of domestic property classes for its application to rural housing markets, and to investigate the interrelationship between domestic property classes and housing consumption cleavages.

Conclusion

Finally, remembering once again the stated objectives of this book, it has been found that local authorities have been given conflicting objectives by central government, and that their attempts to resolve their dilemmas and also to mediate between the central state and local people have not been successful, largely because of the constraints imposed upon them by central government. Restrictive development control policies have protected landscape from housebuilding at the cost of regressive distributional consequences which favour propertied interests in rural areas and exacerbate the disadvantages faced by lower-income groups. Financial restrictions on local authority housebuilding and other forms of social housing provision have led to a dominance of owner-occupied housing and an allocation system based on ability to pay, except in the crofting areas of Scotland. Even there, a clear economic basis to rural housing disadvantage exists, as well as a relationship to life-cycle.

The analysis of the forces which produce the financial and political constraints which in turn prevent local authorities resolving these dilemmas and ameliorating rural housing disadvantage was always outside the scope of this volume, and has been left for future research. However, the consequences of public intervention on various consumers of housing have been identified, and alternative policies have been discussed which might reduce rural housing disadvantage while also giving weight to landscape protection.

Inevitably an agenda for future research has also become apparent.

One item for inclusion on that agenda should be an attempt to explain why such alternative policies will not be implemented and so to understand why rural housing disadvantages will persist.

Bibliography

ACC (1979) *Rural Deprivation*, London: Association of County Councils.

Alexander, D. (1987) 'Shelter's rural housing initiative', in B. MacGregor, D. Robertson and M. Shucksmith (eds) *Rural Housing: Recent Research and Policy*, Aberdeen University Press.

Alexander, D. and Armstrong, N. (1984) 'Housing renewal in Scotland', *The Planner*, July, 26–7.

Allerdale District Council (1978) *Housing Review 1978*, Whitehaven: Allerdale District Council.

Alonso, W. (1960) 'A theory of the urban land market', *Papers and Proceedings of the Regional Science Association*, 6, 149–57.

Alonso, W. (1964) *Location and Land Use*, Cambridge, Mass.: Harvard University Press.

Ambrose, P. (1974) *The Quiet Revolution*, London: Chatto and Windus.

Ambrose, P. (1986) *Whatever Happened to Planning*? London: Methuen.

Ambrose, P. and Colenutt, R. (1975) *The Property Machine*, Harmondsworth: Penguin.

Argyll and Bute District Council (1979) *Housing Plan 1980–85*, Lochgilphead: Argyll and Bute District Council.

Ashton, J., Buckwell, A.E., Harvey, D.R., Thomson, K.J. and Whitby, M.C. (1979) *Measuring the Cost of the CAP*, Dept. of Agricultural Economics, University of Newcastle upon Tyne.

Balchin, P. (1985) *Housing Policy: An Introduction*, London: Croom Helm.

Ball, M. (1983) *Housing Policy and Economic Power*, London: Methuen.

Ball, M. (1986) *Home Ownership: A Suitable Case for Reform*, London: Shelter.

Ball, M. (1988) *Rebuilding Construction: Economic Change in the British Construction Industry*, London: Routledge.

Banff and Buchan District Council (1981) *Housing Plan 1980–85*, Peterhead: Banff and Buchan District Council.

Barlow, J. (1988) 'The politics of land use into the 1990s: landowners, developers and farmers in lowland Britain', *Policy and Politics*, 16 (2), 111–21.

Barlow Report (1940) *Report of the Royal Commission on the Distribution of the Industrial Population*, London: HMSO.

Barr, J. (1967) 'A two-home democracy?', *New Society*, 7 September, 313–15.
Bassett, K. and Short, J. (1980) *Housing and Residential Structure: Alternative Approaches*, London: Routledge & Kegan Paul.
Baum, A. (1982) *Statutory Valuations*, London: Routledge & Kegan Paul.
Baumol, W. and Oates, W. (1975) *The Theory of Environmental Policy*, Englewood Cliffs: Prentice Hall.
BBC (1979) Grapevine: Rural factpack, 15 December.
Beazley, M., Gavin, D., Gillon, S., Raine, C. and Staunton, M. (1980) *The Sale of Council Houses in a Rural Area*, Working Paper 44, Department of Town Planning, Oxford Polytechnic.
Bell, M. (1987) 'The future use of agricultural land in the UK', paper given at the Agricultural Economics Society's Annual Conference.
Bell, P. and Cloke, P. (1989) 'The changing relationship between the private and public sectors: privatisation and rural Britain', *Journal of Rural Studies*, 5 (1), 1–16.
Bennet, A. (1987) 'Progress on new settlements', *The Planner*, November, 35.
Bennett, S. (1975) 'School leaver surveys: Patterdale and Coniston', unpublished report, available on file in LDSPB offices, Kendal.
Bennett, S. (1976) *Rural Housing in the Lake District*, Lancaster University.
Bennett, S. (1977) 'Housing need and the rural housing market', in G. Williams (ed.) *Community Development in Countryside Planning*, Department of Town and Country Planning, Manchester University.
Berry, G. and Beard, G. (1980) *The Lake District: A Century of Conservation*, Edinburgh: Bartholomew.
Best, R. (1981) *Land Use and Living Space*, London: Methuen.
Bielckus, C., Rogers, A.W. and Wibberley, G.P. (1972) *Second Homes in England and Wales*, Wye College, University of London.
Blunden, J. and Curry, N. (1985) *The Changing Countryside*, London: Croom Helm/Open University.
Boddy, M. (1976) The structure of mortgage finance: building societies in the British social formation', *Transactions of the Institute of British Geographers*, 1 (1), 58–71.
Boddy, M. (1980) *The Building Societies*, London: Macmillan.
Bond, A. (1985) 'Housing problems in rural Cumbria', unpublished Open University dissertation.
Bradshaw, J. (1972) 'A taxonomy of social need', in G. McLachlan (ed.) *Problems and Progress in Medical Care*, Seventh Series, Oxford: Oxford University Press.
Bowers, J.K. and Cheshire, P. (1983) *Agriculture, the Countryside and Land Use: an Economic Critique*, London: Methuen.
Bramley, G. (1989) *Meeting Housing Needs*, School of Advanced Urban Studies, University of Bristol.
Bramley, G. and Paice, D. (1987) 'Housing needs in non-metropolitan areas', report of research carried out for the Association of District Councils, Bristol: SAUS.
Brotherton, I. (1981) *Conflict, Consensus, Concern and the Administration of Britain's National Parks*, Department of Landscape Architecture, Sheffield University.

Brown, H. J. (1974) Correspondence, *Journal of Planning and Environmental Law*, 508–9.
Bruce, C. (1986) 'Housing for the elderly in rural communities', unpublished M Phil. dissertation, Edinburgh University.
Bryant, J. (1980) 'Housing for the poor in Papua New Guinea: self help as the only future', in R. Jackson (ed.) *An Introduction to the Urban Geography of Papua New Guinea*, Port Moresby: University of Papua New Guinea.
Buchanan, S. (1982) 'Power and planning in rural areas: preparation of the Suffolk County Structure Plan', in M.J. Moseley (ed.) *Power, Planning and People in Rural East Anglia*.
Buckinghamshire County Council (1977) Structure Plan, written statement, Rickmansworth: Buckinghamshire County Council.
Buckley, M. (1987) *Farm Output Loss in Land Use Planning Decisions*, Occasional Paper 2/87, Dept of Town Planning, South Bank Polytechnic, London.
Bucknall, B. (1984) *Housing Finance*, London: CIPFA
Buller, H. and Lowe, P. (1982) 'Politics and class in rural preservation: a study of the Suffolk Preservation Society', in M.J. Moseley (ed.) *Power, Planning and People in Rural East Anglia*.
Burbridge, V. and Robertson, J. (1978) *The Rural Indicators Study*, Edinburgh: Scottish Office Central Research Unit.
Burney, E. (1967) *Housing on Trial*, London: Oxford University Press.
Campbell, D. (1985)'The Real Crisis of Scottish Agriculture', *Scottish Government Yearbook 1985*, 107–33.
Capstick, M. (1972) *Some Aspects of the Economic Effects of Tourism in the Westmorland Lake District*, Lancaster University.
Capstick, M. (1977) 'Economic, social and political structures in the uplands of Cumbria', in R.B. Tranter (ed.), *The Future of Upland Britain*, CAS, Reading University.
Capstick, M. (1980) *A Critical Appraisal of the Operation of Three Ad Hoc Planning Authorities*, Centre for North West Regional Studies, University of Lancaster.
Capstick, M. (1987) *Housing Dilemmas in the Lake District*, Centre for North West Regional Studies, University of Lancaster.
Capstick, M. and Halliday, G. (1977) 'Social engineering v. social participation in the Lake District National Park: a comment', *University of Lancaster Regional Bulletin*, Spring, 1–4.
CAS (Centre for Agricultural Strategy) (1986) 'Countryside implications for England and Wales of possible changes in the CAP', Main report to DoE and Development Commission, CAS, Reading.
Castells, M. (1977) *The Urban Question*, London: Edward Arnold.
Charles, S. (1977) *Housing Economics*, London: Macmillan
Charles, S. (1978) 'Do we need a housing policy?', *National Westminster Quarterly Review*, August, 57–64.
Charles, S. (1979) *Housing Economics*, London Macmillan.
Claridge, C. (1987) 'Development in the countryside', in B. MacGregor, D. Robertson and M. Shucksmith (eds) *Rural Housing: Recent Research and Policy*.

Clark, D. (1980) *Rural Housing in East Hampshire*, London: National Council for Voluntary Organisation.
Clark, D. (1981) *Rural Housing Initiatives Paper 4: Self-Building Schemes in the Countryside*, London: NCVO.
Clark, D. (1982) *Rural Housing Initiatives*, Proceedings of Seminar D, PTRC Annual Summer Conference, London: PTRC.
Clark, D. (1988) *Affordable Houses in the Countryside: a Role for Private Builders*, ACRE and the House Builders' Federation.
Clark, G. (1982) *Housing and Planning in the Countryside*, Chichester: John Wiley.
Clarke, S. (1982) *Marx, Marginalism and Modern Sociology: from Adam Smith to Max Weber*, London: Macmillan.
Cloke, P. (1977) 'An index of rurality for England and Wales', *Regional Studies*, 11, 34–46.
Cloke, P. (1979) *Key Settlements in Rural Areas*, London: Methuen.
Cloke, P. (1983) *An Introduction to Rural Settlement Planning*, London: Methuen.
Cloke, P. (ed.) (1987) *Rural Planning: Policy into Action*?, London: Harper & Row.
Cloke, P. and Little, J. (1987) 'Policy planning and the State in the rural localities', *Journal of Rural Studies*, 3(4), 343–51.
Cloke, P. and Woodward, N. (1981) 'Methodological problems in the economic evaluation of rural settlement planning', in N. Curry (ed.) *Rural settlement policy and economics, Gloucestershire Papers in Local and Rural Planning*, 12.
Coles, R. (1982) 'Retirement in rural north Norfolk', in M. J. Moseley (ed.) *Power, Planning and People in Rural East Anglia.*
Constable, M. (1988) 'Update on section 52 agreements and local needs statements', *Village Housing*, 3.
Couper, M. and Brindley, T. (1975) 'Housing classes and housing values', *Sociological Review*, 23, 563–76.
Craven, E. (1975) 'Private residential expansion in Kent', in R. E. Pahl (ed.) *Whose City? And Further Essays on Urban Society.*
Croft, P. (1988) 'The New Forest experience', speech at HCT seminar, London.
Crofters' Commission (1967) *Annual Report 1967*, Inverness: Crofters' Commission.
Crofters' Commission (1981) *Annual Report 1981*. Inverness: Crofters' Commission.
Culyer, A. J. (1976) *Need and the National Health Service*, London: Martin Robertson.
Cumbria Countryside Conference (1979) *Rural Housing*, Carlisle: Cumbria Countryside Conference.
Cumbria County Council (1975) 'Technical report on key issues for the Cumbria and Lake District Joint Structure Plan', Carlisle: Cumbria County Council.
Cumbria County Council (1976) 'Cumbria and Lake District Joint Structure Plan', report of survey, Carlisle: Cumbria County Council.

Cumbria County Council (1980) 'Cumbria and Lake District Joint Structure Plan', written statement, Carlisle: Cumbria County Council/Kendal: LDSPB.

DAFS (nd) CBGLS handbook, Edinburgh: DAFS.

DAFS (1979) Appendix to Chapter 6 in SDD (1979) *Housing in Rural Scotland*, unpublished report.

Davies, B. (1968) *Social Needs and Resources in Local Services*, London: Michael Joseph.

Davies, B. (1978) *'Issues on need'*, in R. Davies and P. Hall (eds) *Issues in Urban Society*, Harmondsworth: Penguin.

Davies, J. and Taylor, J. (1970) 'Race, community and no conflict', *New Society*, 9, 67–9.

Dennis, N. (1972) *Public Participation and Planners' Blight*, London: Faber and Faber.

Derounian, J. (1979) *Structure Plans and Rural Communities* London: NCVO.

Derounian, J. (1980) 'The impact of structure plans on rural communities', *The Planner*, July, 87.

Dickens, P., Duncan, S., Goodwin, M. and Gray, F. (1985) *Housing, States and Localities*, London: Methuen.

DoE (1976) 'Circular 4/76: Report of the National Park Policies Review Committee', London: DoE.

DoE (1978) 'Housing in National Parks', minutes of a meeting between LDSPB and DoE on 28 April 1978, unpublished.

DoE (1981a) Proposed modifications to the Cumbria and Lake District Joint Structure Plan.

DoE (1981b) 'Cumbria and Lake District Joint Structure Plan', Examination in Public: Report of the Panel.

DoE (1983) Modifications to the Cumbria and Lake District Joint Structure Plan.

DoE (1987) 'Development involving agricultural land, Circular 16/87', London: DoE.

DoE (1988a) 'Housing in Rural Areas: A Statement by the Secretary of State', London: DoE.

DoE (1988b) 'Housing in Rural Areas: Village Housing and New Villages: A Discussion Paper; London: DoE.

Drewett, R. (1973) 'Land values and the suburban land market', in P. Hall *et al.* (1973) *The Containment of Urban England*.

Duncan, S. S. (1977) 'Alienation and explanation in Human Geography', Graduate Geography School, London School of Economics, Discussion Paper 63.

Duncan, S. S. (1983) 'State intervention and efficient capitalism: separating land ownership from land development in Sweden', paper to Oxford Land Policy Conference, later published in S. Barrett and P. Healey (eds) (1985) *Land Policy: Problems and Alternatives*, Aldershot: Gower.

Duncan, T. (1985) *Rural Housing Initiatives in Scotland*, Glasgow: Planning Exchange.

Duncan, T. (1987) 'House conditions in rural areas: a commentary', in B.

MacGregor, D. Robertson and M. Shucksmith (eds) *Rural Housing: Recent Research and Policy*.

Dunipace, R. (1988) 'The Country Landowners Association Rural Housing Discussion Paper' in M. Winter and A. Rogers (eds) *Who Can Afford to Live in the Countryside? Access to Housing Land*, Royal Agricultural College, Centre for Rural Studies.

Dunleavy, P. (1979) 'The urban bases of political alignment', *British Journal of Political Science*, 9, 409–43.

Dunn, M., Rawson, M. and Rogers, A. W. (1981) *Rural Housing: Competition and Choice*, London: Allen & Unwin.

Edel, M. (1982) 'Home ownership and working class unity', *International Journal of Urban and Regional Research*, 6, 205–22

Eden District Council (1951) *House Building Programme*, Penrith: Eden District Council.

Edwards, A. (1986) 'An agricultural land budget for the UK', Wye College, London.

Elliott, B. and McCrone, D. (1975) 'Landlords as urban managers: a dissenting opinion', in M. Harloe (ed.) *Proceedings of the Conference on Urban Change and Conflict*, London: Centre for Environmental Studies.

Elson, M. (1981) 'Structure plan policies for pressured rural areas', *Countryside Planning Yearbook*, 2, 49–70.

Elson, M. (1986) *Green Belts, Conflict and Mediation in the Urban Fringe* London: Hutchinson.

Elson, M., Gault, I. and Healey, P. (eds) (1979) *Local Needs in Areas of Restraint*, Working Paper No. 42, Department of Town Planning, Oxford Polytechnic.

English, M. and Martin F. M. (eds) (1983) *Social Services in Scotland* Edinburgh: Scottish Academic Press.

Ennew, J. (1980) *The Western Isles Today*, Cambridge: Cambridge University Press.

Evans, A. (1983) 'The determination of the price of land', *Urban Studies*, 20, 119–29.

Evans, A. (1987) *Land Availability for Housing in the South East*, London: House Builders' Federation.

Farming and the Nation (1979) Cmnd. 7458, London: HMSO.

Feist, M., Leat, P., and Wibberley, G. P. (1976) *A Study of the Hartsop Valley*, Cheltenham: Countryside Commission.

Folkesdotter, G. (1987) 'Research and policy for rural housing in Sweden', in B. MacGregor, D. Robertson and M. Shucksmith (eds) *Rural Housing: Recent Research and Policy*.

Food From Our Own Resources (1975) Cmnd. 6020, London: HMSO.

Fordham, R. C. (1971) 'Turning our farms into gardens', *New Society*, 8, July.

Forsythe, D. (1980) 'Urban incomers and rural change', *Sociologia Ruralis*, 20(4), 287–307.

Foulis, M. (1987) 'The effect of sales on the public sector in Scotland', in D. Clapham and J. English (eds) *Public Housing: Current Trends and Future Developments*, London: Croom Helm.

Friend, A. (1980) *A Giant Step Backwards: Council House Sales and Housing Policy*, Occasional Paper 5, London: Catholic Housing Aid Society.

Frith, D and Matthews, J. A. (1974) *Expanding Rural Communities in West Cumbria; Final Report to the Development Commission*, Carlisle: Cumbria Council for Voluntary Action.

Garner, J. F. (1974) Correspondence, *Journal of Planning and Environmental Law*, 511.

Gasson, R. (1975) *Provision of Tied Cottages*, Occasional Paper 4, Department of Land Economy, University of Cambridge.

Gibb, A. and MacLennan, D. (1986) 'Policy and process in Scottish housing, 1950 to 1980', in R. Saville (ed.) *The Economic History of Modern Scotland*, Edinburgh: John Donald.

Gilder, I. M. (1979) 'Rural planning policies: an economic appraisal', *Progress in Planning*, 11 (3), 213–71.

Gilder, I. M. and McLaughlin, B. P. (1978) *Rural Communities in West Suffolk*, Chelmsford: Chelmer Institute of Higher Education.

Gillett, E. (1983) *Investment in the Environment*, Aberdeen: Aberdeen University Press.

Goldsmith, M. (1980) *Politics, Planning and the City*, London: Hutchinson.

Gordon District Council (1978) *Housing Plan 1978–83*, Inverurie: GDC.

Gould, L. (1986) *Changes in Land Use in England, Wales and Scotland, 1985 to 2000 and 2025: Final Report*. Peterborough: Nature Conservancy Council.

Grant, M. (1982) *Urban Planning Law*, London: Sweet and Maxwell.

Gray, F. (1976) 'Selection and allocation in council housing', *Transactions of the Institute of British Geographers*, 1, 34–46.

Greenwood, J. L. (1989) *Planning for Low Cost Rural Housing*, Oxford Polytechnic, School of Planning, Working Paper 112.

Grigson, S. (1986) *House Prices in Perspective – A Review of South East Evidence*, Paper to London and South East Regional Conference.

Haddon, R. (1970) 'A minority in a welfare state society', *New Atlantis*, 2, 80–133.

Hahn, F. (1973) *On the Notion of Equilibrium in Economics*, Cambridge: Cambridge University Press.

Hall, P., Gracey, H., Drewitt, R. and Thomas, R. (1973) *The Containment of Urban England*, London: Allen & Unwin.

Hallett, G. (1977) *Housing and Land Policies in Britain and West Germany*, London: Macmillan.

Hamnett, C. (1976) *Inequality Within Nations: Inequality in Housing*, Milton Keynes: Open University.

Harbison, J. (1983) *Temporary Accommodation Survey*, Portree: Skye and Lochalsh District Council.

Harper, S. (1987a) 'The rural–urban interface in England: a framework of analysis', *Transactions of the Institute of British Geographers*, (New Series), 12 (3), 284–302.

Harper, S. (1987b) 'A humanistic approach to the study of rural populations', *Journal of Rural Studies*, 3.

Harvey, D. (1973) *Social Justice and the City*, London: Edward Arnold.

Harvey, D. (1974) 'Class monopoly rent, finance capital and the urban revolution', *Regional Studies*, 8, 239–55.
Harvey, D. (1977) 'Government policies, financial institutions and neighbourhood change in United States cities', in M. Harloe (ed.) *Capitive Cities*, London: John Wiley.
Heraud, B. J. (1968) 'Social class and the new towns', *Urban Studies*, 5, 33–58.
Herington, J. M. (1984) *The Outer City*, London: Harper & Row.
Herington, J. M. (1986) 'Exurban housing mobility: the implications for future study', *Journal of Economic and Social Geography*, 77(3), 178–86.
Herington, J. M. (1987) 'The centralisation of physical planning decisions in Britain', Paper presented at Anglo-Swedish seminar, Gävle.
HIDB (1974a) *Rural Housing in the Highlands and Islands*, Inverness: HIDB.
HIDB (1974b) *Housing Improvement Survey, Barra 1972*, Inverness: HIDB.
Highland Perthsire Community Councils Forum (1987) *Rural Housing: Report of a Symposium on Assessing Housing Need*, Perth: SCVO.
HMSO (1971) *A Fair Deal for Housing*, London: Department of the Environment, Cmnd. 4728.
HMSO (1974) *Report of the National Park Policies Review Committee*, (The Sandford Report), London: HMSO.
HMSO (1976) *National Parks and the Countryside*, Sixth Report of the House of Commons Expenditure Committee, 1975/76.
HMSO (1977) *Housing Policy, A Consultative Document*, London: Department of the Environment, Cmnd. 6851.
HM Treasury (1976) *Rural Depopulation*, report of an inter-departmental group.
Hodge, I. D. (1984) 'Uncertainty, irreversibility and the loss of agricultural land', *Journal of Agricultural Economics*, 35, 191–202.
Hodge, I. D. and Whitby, M. C. (1981) *Rural Employment: Trends, Options, Choices*, London: Methuen.
Hoggart, K. and Buller, H. (1987) *Rural Development: A Geographical Perspective*, London: Croom Helm.
Hooper, A., Finch, P. and Rogers, S. (1988) 'Land supply in the shire counties', in M. Winter and A. Rogers (eds) *Who Can Afford to Live in the Countryside? Access to Housing Land.*
House Builders' Federation (1987) response to DoE draft circular on development involving agricultural land, as reported in the *Guardian*, 10 February.
House Builders' Federation (1988) *Homes in the Countryside: A New Partnership*.
Housing Corporation (1985) Annual Report 1984–85.
Hughes, J. (1987) 'Housing and rural development', in B. MacGregor, D. Robertson and M. Shucksmith (eds) *Rural Housing: Recent Research and Policy*.
Hunter, J. (1976) *The Making of the Crofting Community*, Edinburgh: John Donald.
Hunter, J. (1980) 'When a threat hangs over home sweet home', *Press and Journal*, 17 March.

Hunter, J. (1986) *Skye – The Island*, Edinburgh: Mainstream.
Jacobs, C. (1972) *Second Homes in Denbighshire*, Denbighshire County Council.
Jennings, R. (1987) 'Rural housing and housing policy in Ireland', in B. MacGregor, D. Robertson and M. Shucksmith (eds) *Rural Housing: Recent Research and Policy*.
JLRC (1983) *Is there sufficient housing land for the 1980s? Paper II: How many houses have we planned for: Is there a problem*?, London: JLRC/HBF.
JLRC (1984) *Housing and Land 1984-1991. 1992-2000: How Many Houses Will We Build? What Will Be the Effect on our Countryside?*, London: JLRC/HBF.
Johnston, W. and Partners (1979) Quantity Surveyors' Letter to the Chief Architect, South Lakeland District Council, concerning the extra cost of building houses in the LDSPB area, 8 January.
Jones, H. (1987) 'Incomers to peripheral rural areas in northern Scotland', in B. MacGregor, D. Robertson and M. Shucksmith (eds) *Rural Housing: Recent Research and Policy*.
Jones, H., Caird, J., Berry, W. and Dewhurst, J. (1986) 'Peripheral counter-urbanisation: findings from an integration of census and survey data in northern Scotland', *Regional Studies*, 20 (1), 15–26.
Karn, V. (1976) *Priorities for Local Authority Mortgage Lending: A Case Study of Birmingham*, Research Memorandum 52, Centre for Urban and Regional Studies, University of Birmingham.
Kemp, P. (1989) 'Why council housing?, *Roof*, 14 (2), 42–4.
Kilroy, B. (1982) 'The financial and economic implications of council house sales', in J. English (ed.) *The Future of Council Housing*, London: Croom Helm.
King, D. (1987) 'Grasping the nettle in the numbers game', *Planning*, 747, 10–11.
Kinghan M. (1981) 'Create your own housing', *The Planner*, 67 (3), 68–9.
Kramer, J. and Young, K. (1978) *Strategy and Conflict in Metropolitan Housing*, London: Heinemann.
LDPB (1972) Minutes, Finance and General Purposes Committee, 6 June.
LDPB (1973) Minutes, Finance and General Purposes Committee, 4 September.
LDSPB (1974a) 22nd Annual Report, 1973–74, Kendal: LDSPB.
LDSPB (1974b) Minutes, Plans and Policy Sub-Committee, 8 May.
LDSPB (1974c) Minutes, Planning Committee, 9 September.
LDSPB (1975) 23rd Annual Report 1974–75, Kendal: LDSPB.
LDSPB (1976) 24th Annual Report 1975–76, Kendal: LDSPB.
LDSPB (1977a) 25th Annual Report 1976–77, Kendal: LDSPB.
LDSPB (1977b) Draft National Park Plan, Kendal: LDSPB.
LDSPB (1977c) Minutes, Development Control Committee, 5 September.
LDSPB (1977d) Minutes, Planning Committee, 2 November.
LDSPB (1977e) Minutes, LDSPB full meeting, 24 November.
LDSPB (1978a) 26th Annual Report 1977–78, Kendal: LDSPB.

LDSPB (1978b) National Park Plan, Kendal: LDSPB.
LDSPB (1979) 27th Annual Report 1978–79, Kendal: LDSPB.
LDSPB (1980a) Cumbria and Lake District Joint Structure Plan, written statement, Kendal: LDSPB/Carlisle: Cumbria County Council.
LDSPB (1980b) Supplementary statement no.1 to examination in public of structure plan, Kendal: LDSPB.
LDSPB (1982) 30th Annual Report 1981–82, Kendal: LDSPB.
LDSPB (1985) Draft National Park Plan Review, Kendal: LDSPB.
LDSPB (1986) Draft Keswick Local Plan, Kendal: LDSPB.
LDSPB (1989) Annual Report 1988–89, Kendal: LDSPB.
Lambert, C. (1976) *Building Societies, Surveyors and the Older Areas of Birmingham*, Working Paper 38, Centre for Urban and Regional Studies, University of Birmingham.
Lansley, S. (1979) *Housing and Public Policy*, London: Croom Helm.
Larkin, A. (1978a) 'Housing and the poor', in A. Walker (ed.) *Rural Poverty: Poverty, Deprivation and Planning In Rural Areas*, London: Child Poverty Action Group.
Larkin, A. (1978b) 'Rural housing – too dear, too few and too far', *Roof*, January, 15–17.
Larkin, A. (1978c) 'Second homes – their position in rural housing provision', in M. Talbot and R. W. Vickerman (eds) *Proceedings of a Conference of the Leisure Studies Association*, Conference Paper 8.
Larkin, A. (1979) 'Rural housing and housing needs', in J. M. Shaw (ed.) *Rural Deprivation and Planning*, Norwich: Geo. Books.
Lean, G. (1989) 'The destruction of the British countryside', a four-part series in *The Observer* colour magazine, June/July.
Lewis Council of Social Services (1986) *Community Insulation Project East and West Lewis: An Assessment of Need*, Stornoway: Lewis CSS.
Lichfield, N. and Darin-Drabkin, H. (1980) *Land Policy in Planning*, London, Allen & Unwin.
Lock, D. (1989) 'Green belt bargain', *Roof*, 14 (3), 42–3.
Loughlin, M. (1984) *Local Needs Policies and Development Control Strategies*, SAUS Working Paper 42, Bristol University.
Lowe, P. (1988) 'The politics of land', in M. Winter and A. Rogers *Who can afford to live in the countryside? Access to housing land.*
Lowe, P. and Winter, M. (1987) 'Alternative perspectives on the alternative land use debate', in N. R. Jenkins and M. Bell (eds) *Farm Extensification: Implications of EC Regulation 1760/87*, ITE: Merlewood Research and Development Paper 112.
Lumb, R. (1980) *Recent Migration in the Highlands and Islands*, Aberdeen, ISSPA.
McDowell, L. (1982) 'Urban housing markets', Unit 12, Block 3, in *Open University Course D202 – Urban Change and Conflict*, Milton Keynes: Open University Press.
MacEwen, A. and MacEwen, M. (1982) *National Parks: Conservation or Cosmetics?*, London: Allen & Unwin.
MacEwen, M. and Sinclair, G. (1983) *New Life for the Hills*, London: Council for National Parks.

MacGregor, B., Robertson, D. and Shucksmith, M. (eds) (1987) *Rural Housing: Recent Research and Policy*, Aberdeen: Aberdeen University Press.
Mackay, G. A. and Laing, G. (1982) *Consumer Problems in Rural Areas*, Glasgow: Scottish Consumer Council.
Mackay, J. (1987) 'Crofter Housing', minister's response to Scottish Crofters' Union policy statement.
McLaughlin, B. P. (1980) 'Rural deprivation', *The Planner*, 67 (2), March, 31–3.
McLaughlin, B. P. (1985) 'Assessing the extent of rural deprivation', *Journal of Agricultural Economics*, 26 (1), 77–80.
McLaughlin, B. P. (1986) 'The rhetoric and the reality of rural deprivation', *Journal of Rural Studies*, 2 (4), 291–308.
MacLennan, D. (1982a) *Housing Economics: An Applied Approach*, London: Longmans.
MacLennan, D. (1982b) 'Public cuts and private sector slump: Scottish housing policy in the 1980s', in M. Cuthbert (ed.) *Government Spending in Scotland*, Edinburgh: Paul Harris.
MacLennan, D. (1986) *The Demand for Housing: Economic Perspectives and Planning Practices*, Edinburgh: SDD.
Malpass, P. and Murie, A. (1987) *Housing Policy and Practice* (2nd edition), London: Macmillan.
Martin, T. (1988) 'The dimensions of housing need in rural Scotland' in T. Martin and J. Doherty (eds) *The Nature of the Scottish Housing Crisis*, Edinburgh: Royal Scottish Geographical Society.
Martin and Voorhees Associates (1980) *Review of Rural Settlement Policies*, Report to DoE, SDD and Welsh Office.
Merrett, S. (1982) *Owner Occupation in Britain*, London: Routledge & Kegan Paul.
Middleton, R. (1988) 'Community needs housing', in M. Winter and A. Rogers *Who Can Afford to Live in the Countryside? Access to Housing Land*.
Ministry of Housing and Local Government (1960) Development Control Policy Note 4, London: HMSO.
Moore, V. (1974) Correspondence, *Journal of Planning and Environmental Law*, 513.
Morcombe, K. (1984) *The Residential Development Process*, Farnborough: Gower.
Moseley, M. (1980) 'Is rural deprivation really rural?', *The Planner*, July, 97.
Moseley, M. (1981) *Rural Development and its Relevance to the Inner City Debate*, London: Social Science Research Council.
Moseley, M. (ed.) (1982) *Power, Planning and People in Rural East Anglia*, Norwich: Centre for East Anglian Studies, University of East Anglia.
Moser, C. A. and Kalton, G. (1971) *Survey Methods in Social Investigation*, London: Heinemann.
Munton, R. and Goodchild, B. (1985) *Development and the Landowner*, London: Allen & Unwin.

Murie, A. (1976) 'Estimating housing need: technique or mystique?', *Housing Review*, May–June, 54–8.
Murie, A., Niner, P. and Watson, C. (1977) *Housing Policy and the Housing System*, London: Allen & Unwin.
Nath, S. K. (1973) *A Perspective of Welfare Economics*, London: Macmillan.
National Agricultural Centre Housing Association (1978) *Rural Housing Today and Tomorrow*, Stoneleigh: NACHA.
National Agricultural Centre Rural Trust (1987) *Village Homes for Village People*, London: NACRT.
National Agricultural Centre Rural Trust (1988) *Village Housing*, Issues 2 and 3.
National Agricultural Centre Rural Trust (1989) *Guide to Village Housing*, London: NACRT.
National Trust (1981) *A Strategy Plan for the National Trust in the Lake District*, London: National Trust.
Needleman, L. (1964) *The Economics of Housing*, London: Staples Press.
Neuberger, H. and Nicol, B. (1975) *The Recent Course of Land and Property Prices and the Factors Underlying It*, London: Department of the Environment.
Newby, H. (1977) *The Deferential Worker*, London: Allen Lane.
Newby, H. (1980) *Green and Pleasant Land? Social Change in Rural England*, Harmondsworth: Penguin.
Newby, H. (1981) 'Urbanism and the rural class structure', in M. Harloe (ed.) *New Perspectives in Urban Change and Conflict*, London: Heinemann.
Newby, H. (1982) 'Rural sociology and its relevance to the agricultural economist: a review', *Journal of Agricultural Economics*, 32, 125–65.
Newby, H., Bell, C., Rose, D., and Saunders, P. (1978) *Property, Paternalism and Power: Class and Control in Rural England*, London: Hutchinson.
Newby, H., Bell. C., Saunders, P. and Rose, D. (1977) 'Farmers' attitudes to conservation', *Countryside Recreation Review*, 2, 23–30.
Newby, H. *et al.* (1981) 'Farming for survival: the small farmer in the contemporary rural class structure', in F. Bechofer and B. Elliot (eds) *The Petit Bourgeoisie*, London: Macmillan.
NFHA (1982) *Rural Housing: Hidden Problems and Possible Solutions*, London: National Federation of Housing Associations.
Niner, P. (1975) *Local Authority Housing Policy and Practice*, Occasional Paper 31, Centre for Urban and Regional Studies, University of Birmingham.
Niner, P. (1978) *Homes to Let*, Occasional Paper, Community Development Programme Unit, University of York.
Noble, D. H. (1986) Discussion Paper on Rural Housing Initiatives, report to Skye and Lochalsh Housing Committee, 23 June.
Norfolk County Council (1977) Structure Plan, Written Statement, Norwich: Norfolk County Council.
North, J. (1987) in M. Bell 'The future use of agricultural land in the UK'.
North Wiltshire District Council (1977) Short Term Leasing Scheme, internal report, 16 December.

Northern Region Strategy Team (1976) *Housing in the Northern Region*, report no. 15 and technical appendices, Newcastle upon Tyne: NRST.
Nuffield Foundation (1986) *Town and Country Planning*, London: Nuffield Foundation.
Pacione, M. (1984) *Rural Geography*, London: Harper & Row.
Pahl, R. E. (1966) 'The social objectives of village planning', *Official Architecture and Planning*, 29(8), 1145–6.
Pahl, R. E. (1975) *Whose City? And Further Essays on Urban Society*, Harmondsworth: Penguin.
Pahl, R. E. (1977) 'Managers, technical experts and the state: forms of mediation, manipulation and dominance in urban and regional development', in M. Harloe (ed.) *Captive Cities: Studies in the Political Economy of Cities and Regions*, Chichester: Wiley.
Pahl, R. E. (1978) 'Castells and collective consumption, *Sociology*, 12, 309–15.
Pahl, R. E. (1979) 'Socio-political factors in resource allocation', in D. T. Hebert and D. M. Smith (eds), *Social Problems and the City*, London: Oxford University Press.
Palmer, R. (1955) 'Realtors as social gatekeepers: a study in social control', unpublished PhD thesis, Yale University.
Penfold, S. (1974) *Housing Problems of Local People in Rural Pressure Areas: The Peak District Experience*, Occasional Paper, Department of Town and Regional Planning, Sheffield University.
Phillips, D. and Williams, A. (1981) 'Council house sales and village life', *New Society*, 58, 993.
Phillips, D. and Williams, A. (1982) *Rural Housing and the Public Sector*, Aldershot: Gower.
Phillips, D. and Williams, A. (1983) 'The social implications of rural housing policy', *Countryside Planning Yearbook*, 4, 77–102.
Phillips, D. and Williams, A. (1984) *Rural Britain: A Social Geography*, Oxford: Blackwell.
PIEDA (1986) *Land Supply and House Prices in Scotland*, Edinburgh: Scottish Development Department.
Planning Exchange (1979) Report on seminar on Housing Action Areas and improving older housing in the Islands, Highland and Grampian regions, Glasgow: Planning Exchange.
Planning Exchange (1984) *Rural Housing Initiatives in Scotland*, Glasgow: Planning Exchange.
Platt, S. (1987) 'Yes Minister, but...', *Roof*, 12, (1), 23–5.
Pratt, G. (1982) 'Class analysis and urban domestic property: a critical re-examination', *International Journal of Urban and Regional Research*, 6, 481–502.
Price, C. (1978) *Landscape Economics*, London: Macmillan.
Rawson, M. (1978) *Rural Housing and Population in South Oxfordshire District: a Statistical Appraisal From Census Data*, Working Paper 6, Wye College, University of London.
Rawson, M. and Rogers, A. W. (1976) *Rural Housing and Structure Plans*,

Working Paper 1, Wye College, University of London.
Reade, E. (1982) 'The effects of town and country planning in Britain', Unit 23, Block 5, in *Open University Course D202 – Urban Change and Conflict*, Milton Keynes: Open University Press.
Reade, E. (1987) *British Town and Country Planning*, Milton Keynes: Open University Press.
Rex, J. and Moore, R. (1967) *Race, Community and Conflict*, London: Oxford University Press.
RICS (1986) *Better Housing for Britain*, London: RICS.
Robson, B. T. (1975) *Urban Social Areas*, London: Oxford University Press.
Robson, B. T. (1979) 'Housing, empiricism and the state', in D. T. Herbert and D. M. Smith (eds) *Social Problems and the City*, London: Oxford University Press.
Rocke, T. (1987) 'Implementation of rural housing policy', in P. Cloke (ed.) *Rural Planning: Policy into Action*?, London: Harper & Row.
Roger Tym and Partners (1987) *Land Used for Residential Development in the South-East*, Summary Report for DoE and SERPLAN, London: Roger Tym and Partners.
Rogers, A. W. (1976) 'Rural housing', in G. E. Cherry (ed.) *Rural Planning Problems*, London: Leonard Hill.
Rogers, A. W. (1977) 'Second homes in England and Wales', in J. T. Coppock (ed.), *Second Homes: Curse or Blessing*?, Oxford, Pergamon.
Rogers, A. W. (1981) 'Housing in the national parks', *Town and Country Planning*, July/August, 193–5.
Rogers, A. W. (1983) 'Rural housing', in M. Pacione (ed.) *Progress in Rural Geography*, London: Croom Helm.
Rogers, A. W. (1985a) 'Local claims on rural housing', *Town Planning Review*, 56, 367–80.
Rogers, A. W. (1985b) 'Rural housing; an issue in search of a focus', *Journal of Agricultural Economics*, 36 (1), 87–9.
Rogers, A. W. (1987) 'Issues in English rural housing: an assessment and prospect', in B. MacGregor, D. Robertson and M. Shucksmith (eds) *Rural Housing: Recent Research and Policy*.
Rose, D., Saunders, P., Newby, H. and Bell, C. (1979) 'The economic and political basis of rural deprivation: a case study', in J. M. Shaw (ed.) *Rural Deprivation*, Norwich: Geo Abstracts Ltd.
Rossi, H. (1977) *Shaw's Guide to the Rent (Agriculture) Act 1976*, London: Shaw.
Rowley, C. K. and Peacock, A.T. (1975) *Welfare Economics: A Liberal Restatement*, London: Robertson.
Rutgers, P. (1972) 'Country cottages in the Lake District', unpublished undergraduate dissertation, University of Newcastle upon Tyne.
Rydin, Y. (1984) 'The struggle for housing land: a case of confused interests', *Policy and Politics*, 12, 431–46.
Rydin, Y. (1985) 'Residential development and the planning system', *Progress in Planning*, 24, 1–69.
Rydin, Y. (1986) *Housing Land Policy*, Aldershot: Gower.

Sandbach, F. R. (1978) 'The early campaign for a national park in the Lake District', *Transactions of the Institute of British Geographers*, 3 (4), 498–514.
Saunders, P. (1978) 'Domestic property and social class', *International Journal of Urban and Regional Research*, 2, 233–51.
Saunders, P. (1980) *Urban Politics: A Sociological Interpretation*, Harmondsworth: Penguin.
Saunders, P. (1981) *Social Theory and the Urban Question*, London: Hutchinson.
Saunders, P. (1984) 'Beyond housing classes: the sociological significance of private property rights in the means of consumption', *International Journal of Urban and Regional Research*, 8, 202–27.
Saunders, P. (1986) *Social Theory and the Urban Question* (Second Edition), London: Hutchinson.
SCLSERP/HBF (1984) *Housing Land Supply in the South-East*, London: SERPLAN.
Scott Report (1942) *Report of the Committee on Land Utilisation in Rural Areas*, Cmnd. 6378, London: HMSO.
Scottish Crofters' Union (1987) *Crofter Housing: The Way Forward*, Broadford: Scottish Crofters' Union.
SDD (1978) *Assessing Housing Needs: Scottish Housing Handbook 1*, Edinburgh: Scottish Office.
SDD (1979) 'Housing in rural Scotland', unpublished report.
SDD (1983a) 'The condition of the housing stock in Scotland', internal working paper.
SDD (1983b) 'House condition and improvement in rural Scotland', internal working paper.
SDD (1986) 'Response to Rural Forum's June 1985 policy statement on Scotland's rural housing', Edinburgh: Scottish Office.
SDD (1987) *Draft National Planning Guidelines: Agricultural Land*, Edinburgh: Scottish Development Department.
Self, P. and Storing, H. (1962) *The State and the Farmer*, London: Allen & Unwin.
SFHA (1984) 'A medium term plan 1984/5–1986/7', Edinburgh: SFHA.
SFHA (1987) *Scottish Federation News*, April, special issue on rural housing.
SHAC (1967) *Scotland's Older Houses*, Report by the Scottish Housing Advisory Committee, Edinburgh: Scottish Development Department.
Schaffer, F. (1972) *The New Town Story*, London: Paladin.
Shaw, J. M. (1980) 'Rural planning in 1980: an overview', *The Planner*, July, 88–90.
Shelter (1981) *Housing in Scotland*, Edinburgh: Shelter.
Shelter (1982) *Council House Allocation in Scotland*, Edinburgh: Shelter.
Shelter (1985) *Who Needs Empty Houses*?, Edinburgh: Shelter
Shelter/Rural Forum (1988) *Scotland's Rural Housing: Opportunities for Action*, Report by D. Alexander, M. Shucksmith and N. Lindsay, Perth: Rural Forum.
Short, J. R. (1979) 'Landlords and the private rented sector: a case study', in

M. Boddy (ed.) *Property, Investment and Land*, Working Paper 2, School for Advanced Urban Study, Bristol University.
Short, J. R. (1982) *Housing in Britain*, London: Methuen.
Short, J. R. Fleming. S. C. and Witt, S. J. G. (1986) *Housebuilding, Planning and Community Action*, London: Routledge & Kegan Paul.
Short, J. R., Witt, S. and Fleming, S. (1987) 'Conflict and compromise in the built environment: housebuilding in central Berkshire', *Transactions of the Institute of British Geographers*, 12, 29–42.
Shucksmith, M. (1980a) 'Local interests in a national park, *Town and Country Planning*, 49, 418–21.
Shucksmith, M. (1980b) 'Housing in the Lake District National Park', Submission to the examination in public of the Cumbria and Lake District Joint Structure Plan.
Shucksmith, M. (1981) *No Homes for Locals?*, Farnborough: Gower.
Shucksmith, M. (1982) *Housing Provision for Lower Income Groups in Rural Areas*, Proceedings of Seminar D, PTRC Annual Summer Conference, London: PTRC.
Shucksmith, M. (1983) 'Second homes: a framework for policy', *Town Planning Review*, 54 (2), 174–93.
Shucksmith, M. (1984) *Scotland's Rural Housing: A Forgotten Problem?*, Perth: Rural Forum.
Shucksmith, M. (1985a) 'Public intervention in rural housing markets', *Planning Outlook*, 28 (2), 70–3.
Shucksmith, M. (1985b) 'Rural housing in Scotland', Paper given to the annual conference of the Rural Economy and Society Study Group of the British Sociological Association, Loughborough University.
Shucksmith, M. (1986) 'Development in the countryside and green belts', *Scottish Planning Law and Practice*, 17, 21.
Shucksmith, M. (1987) *Public Intervention in Rural Housing Markets*, PhD Thesis, University of Newcastle upon Tyne.
Shucksmith, M. (1987a) *An Independent Inquiry into the Crofter Housing Grants and Loans Scheme*, Broadford, Scottish Crofters' Union.
Shucksmith, M. (1987b) 'Rural housing in Scotland', in P. Selman (ed.) *Countryside Planning in Scotland*, Stirling University Press.
Shucksmith, M. (1987c) 'Crofting and island issues', in B. MacGregor, D. Robertson and M. Shucksmith (eds) *Rural Housing: Recent Research and Policy*.
Shucksmith, M. (1987d) 'An analysis of crofter housing policies', Paper given at Ninth International Marginal Regions Seminar, Skye, Proceedings to be published.
Shucksmith, M. (1987e) 'Land use conflicts in the English Lake District', in M. Merlo, G. Stellin, P. Harou and M. C. Whitby (eds) *Multipurpose Agriculture and Forestry*, Kiel: Wissenschaftsverlag Vauk Kiel.
Shucksmith, M. (1988) 'Trends in rural housing provision in Scotland', in J. Doherty and A. Martin (eds) *The Nature of the Scottish Housing Crisis*, Edinburgh University Press.
Shucksmith, M. (1988a) 'Policy aspects of housebuilding on farmland in

Britain', *Land Development Studies*, 5, 129–38.
Shucksmith, M. (1988b) 'Current rural land use issues in Scotland: an overview', *Scottish Geographical Magazine*, 104 (3), 176–80.
Shucksmith, M. (1989) 'Affordable housing in villages', in N. Lindsay (ed.) *Rural Housing in Lincolnshire*, Sleaford: Community Council of Lincolnshire.
Shucksmith, M. (1990) 'An analysis of crofter housing policies', *Scottish Geographical Magazine*, forthcoming.
Shucksmith, M., Alexander, D. and Lindsay, N. (1988) *Scotland's Rural Housing: Opportunities for Action*, Edinburgh: Shelter/Rural Forum.
Shucksmith, M. and Watkins, L. (1988a) *The Conversion of Agricultural Land to Housing: Background to Policy Reform*, Working Paper No. 1, Department of Land Economy, University of Aberdeen.
Shucksmith, M. and Watkins, L. (1988b) *The Conversion of Agricultural Land to Housing: Theoretical Issues*, Working Paper No. 2, Department of Land Economy, University of Aberdeen.
Shucksmith, M. and Watkins, L. (1988c) *The Conversion of Agricultural Land to Housing: Land Prices, House Prices and Land Supply*, Working Paper No. 3, Department of Land Economy, University of Aberdeen.
Shucksmith, M. and Watkins, L. (1988d) *The Conversion of Agricultural Land to Housing: Housing Market Effects*, Working Paper No. 4, Department of Land Economy, University of Aberdeen.
Shucksmith, M. and Watkins, L. (1988e) 'The supply side of rural housing markets', in M. Winter and A. Rogers (eds) *Who Can Afford to Live in the Countryside? Access to Housing Land*.
Shucksmith, M. and Watkins, L. (1990) *The Development of Housing Associations in Rural Scotland*, Research Report 1, Edinburgh: Scottish Homes.
Sim, D. (1983) 'Some changes in new town migration', *Planning Outlook*, 26 (1), 44–50.
Simpson, T. S. (1975) 'Aspects of rural planning with special reference to second homes in the Lake District', unpublished dissertation, University of Newcastle upon Tyne.
Smith, S. (1979) *An Assessment of District Council Housing Plans in Dumfries and Galloway Region*, Edinburgh: Scottish Council for Social Services.
Sokal, R. R. and Michener, C. D. (1958) Kansas University Scientific Bulletin, Volume 38.
South Lakeland District Council (1977) *Housing Investment Programme, Strategy Statement*, Kendal: South Lakeland District Council.
South Lakeland District Council (1980a) *Housing Investment Programme, Strategy Statement 1981/82*, Kendal: South Lakeland District Council.
South Lakeland District Council (1980b) Submission to examination in public of Cumbria and Lake District Joint Structure Plan.
Stewartry District Council (1979) *Housing Plan 1979–84*, Kirkcudbright: Stewartry District Council.
Strathclyde University (1974) *Housing Improvement Surveys Barra 1972*, Inverness: HIDB.

Suffolk County Council (1977) *Structure Plan*, Written Statement, Bury St Edmonds: Suffolk County Council.

Tangermann, S. (1984) 'EEC farm policy: the "reforms" that change nothing', *Financial Times*, 5 September.

Thomas, W. (1987) 'The Inner City Commission', *The Planner*, November, 30–1.

Thomas, C. and Winyard, S. (1979) 'Rural incomes', in J. M. Shaw (ed.) *Rural Deprivation and Planning*, Norwich: Geo Books.

Thompson, F. (1984) *Crofting Years*, Barr: Luath Press.

Thomson, S. (1982) 'New houses in the countryside', *Scottish Planning Law and Practice*, 6, 38–43.

Thunwall, C. (1987) *Stability and Mobility in a Rural Area*, Paper to the British–Swedish International Seminar, Gävle, Sweden.

TRRU (1981) *The Economy of Rural Communities in the National Parks of England and Wales*, TRRU, University of Edinburgh.

Uthwatt Report (1942) *Report of the Expert Committee on Compensation and Betterment*, Cmnd. 6386, London: HMSO.

Vallis, E. (1972) 'Urban land and building prices', parts I–IV, *Estates Gazette*, 222, 1015–19, 1209–13, 1406–7, 1604–5.

Waldegrave, W. (1987) Address to the Planning Inspectorate, reported in *Planning*, 716, 10–11.

Ward, C. (1987) 'An end to public housing?', *Roof*, 12 (1), 13–14.

Watkins, C. and Winter, M. (1988) *Superb Conversions? Farm Diversification – the Farm Building Experience*, Centre for Rural Studies, Royal Agricultural College, Cirencester.

Watkins, L. (1987) 'Improvement policy: the practice of rural authorities', in B. MacGregor, D. Robertson and M. Shucksmith (eds) *Rural Housing: Recent Research and Policy*.

Watkins, L. (1989) *Housing Rehabilitation in Rural Scotland*, PhD Thesis, University of Aberdeen.

Webber, R. and Craig, J. (1978) *Socio-economic Classification of Local Authority Areas*, OPCS Studies on Medical and Population Subjects 35, London: HMSO.

Weber, M. (1968 edn) *Economy and Society*, 3 vols, Berkeley: University of California Press.

Weir, S. (1976) 'Red line districts', *Roof*, July 109–14.

Welsh Office (1982) Response to the memorandum submitted by Gwynedd County Council making proposals for action designed to limit the growth of second homes, February 1982.

Whitby, M. and Willis, K. (1978) *Rural Resource Development: An Economic Approach*, London: Methuen.

White, P. (1986) 'Land availability, land banking and the price of land for housing: a review of recent debates', *Land Development Studies*, 3 (2), 101–11.

Whitehead, C. and Kleinman, M. (1986) *The Future of the Private Rented Sector*, Cambridge, Department of Land Economy, University of Cambridge.

WIIA (1978) *Housing Improvement Surveys, Uist and Barra*, Stornoway: WIIA.

WIIA (1980) Housing Plan 1979–84, Stornoway: WIIA.

WIIA (1985) Housing Plan 1984–89, Stornoway: WIIA.

Williams, A. (1974) 'Need as a demand concept', in A. J. Culyer (ed.) (1974) *Economic Policies and Social Goals: Aspects of Public Choice*, London: Martin Robertson.

Williams, N. and Sewel, J. (1987) 'Council house sales in rural areas', in B. MacGregor, D. Robertson and M. Shucksmith (eds) *Rural Housing: Recent Research and Policy*.

Williams, P. (1976) 'Change in an urban area: the role of institutions in the process of gentrification in the London Borough of Islington', unpublished PhD Thesis, University of Reading.

Williams, P. (1978) 'Urban managerialism: a concept of relevance?', *Area*, 10, 236–40.

Willis, K. (1980) *The Economics of Town and Country Planning*, London: Granada.

Wingo, L. (1961) *Transportation and Urban Land*, Baltimore: John Hopkins Press.

Winter, M. and Rogers, A. (eds) (1988) *Who Can Afford to Live in the Countryside? Access to Housing Land*, Occasional Paper 2, Royal Agricultural College, Centre for Rural Studies.

Winter, H. (1980) *Homes for Locals*, Exeter: Community Council for Devon.

Wishart, D. (1978) *Clustan User Manual* (3rd edn), Report No. 47, Program Library Unit, Edinburgh University.

Wordsworth, W. (1810) *Guide to the Lakes*, Oxford: Oxford University Press.

World Conservation Strategy (1980) Gland: International Union for Conservation of Nature Natural Resources.

Young, E. and Rowan-Robinson, J. (1985) *Scottish Planning Law and Procedure*, Edinburgh: Hodge.

Young, G. M. (1943) *Summary of the Scott and Uthwatt Reports*, Harmondsworth: Penguin.

Young, R. K. (1987) 'Housing associations: their role in the rural context', in B. MacGregor, D. Robertson and M. Shucksmith (eds) *Rural Housing: Recent Research and Policy*.

Index